Sandra Warnecke

Numerical schemes for multi-species BGK equations based on a variational procedure

Sandra Warnecke

Numerical schemes for multi-species BGK equations based on a variational procedure

applied to multi-species BGK equations with velocity-dependent collision frequency and to quantum multi-species BGK equations

Würzburg University Press

Dissertation, Julius-Maximilians-Universität Würzburg
Fakultät für Mathematik und Informatik, 2022
Gutachter: Prof. Dr. Christian Klingenberg, Prof. Dr. Cory Hauck

Impressum

Julius-Maximilians-Universität Würzburg
Würzburg University Press
Universitätsbibliothek Würzburg
Am Hubland
D-97074 Würzburg
www.wup.uni-wuerzburg.de

Print on Demand

Coverdesign: Jürgen Schwarz

ISBN 978-3-95826-192-1 (print)
ISBN 978-3-95826-193-8 (online)
DOI 10.25972/WUP-978-3-95826-193-8
URN urn:nbn:de:bvb:20-opus-282378

Acknowledgments

This PhD project was made possible by financial support of the University of Würzburg giving me a position of a teaching assistant.

I have experienced a lot of help and encouragement from many persons during my dissertation project. Particularly, I would like to thank my supervisor Christian Klingenberg for the constant and friendly support throughout all years of my studies.

I enjoyed and have profited from many discussions with Cory Hauck and Jeffrey Haack. A trip to visit them in the United States was enabled by the financial support of the Bayerische Forschungsallianz (grant no. BaylntAn UWUE 2019-29).

The numerous visitors of our group enriched the daily life and gave insights in many interesting research topics. Among these I especially want to thank Bruno Després for fruitful discussions.

Besides, I am happy to be part of such a lively work group. I particularly thank Marlies Pirner who always answered my many questions and introduced me to the exciting topic of kinetic theory. Personally, I enjoyed the friendly atmosphere in the group with coffee breaks, cake meetings and our joint working trip to Oberwolfach. Thanks to Marlies, Wasilij, Simon, Jonas, Andrea, Marc, Farah, Jayesh, Claudius, Eva, Kathrin, Lena.

I also thank Marlies Pirner and Claudius Birke for carefully reading the thesis draft.

Furthermore, I would like to thank my fellow students and friends. They contributed in various ways to the pleasure and success of my studies.

I am very grateful for my family who supports me in all parts of my life. My final thanks go to my wonderful husband for enriching my life and giving me unconditional support.

Sandra Warnecke

Preface

This PhD thesis arose from a research collaboration between the University of Würzburg and two National Laboratories (Los Alamos National Laboratory and Oak Ridge National Laboratory) in the USA.

The American National Ignition Facility (NIF), which collaborates with the above two national labs, has been working on setting off fusion by inertial confinement fusion. Here one attempts to generate an imploding spherical shock wave that will generate sufficient energy at its center to initiate fusion. Success of this initiative would solve the world's energy needs.

This NIF project has been hampered by technical difficulties. The spherical blast wave has so far not stayed stable when imploding, thus not generating enough energy at its center to start off fusion. One way to make progress are numerical simulations of this process. Part of these simulations are based on models of so-called kinetic partial differential equations. As a prerequisite for this thesis, such a model had been developed in C. Klingenberg, M. Pirner, G. Puppo: 'A consistent kinetic model for a two component mixture with an application to plasma', Kinetic and Related Models Vol. 10, No. 2, pp. 445–465 (2017). This model needed to be extended, which was done in J. Haack, C. Hauck, C. Klingenberg, M. Pirner, S. Warnecke: 'A consistent BGK model with velocity-dependent collision frequency for gas mixtures', Journal of Statistical Physics, Vol. 84, No. 31 (2021). The challenge now was the numerical simulation of this latter model, which is a main contribution of this PhD thesis.

Kinetic models tend to be numerically very costly. The above mentioned models introduce a simplification (BGK kinetic model) that significantly reduces the computational time. Still one had to take care of the fact that not too many of the important physical processes are lost in this simplification.

The success of the new numerical method presented in this thesis can be gauged for example by the successful numerical simulation in Section 5.3.6, something never before achieved with such a simplified BGK kinetic model.

Thus this PhD thesis contributed in a significant way towards the goal of this German-American collaboration.

Würzburg, September 21, 2022
Christian Klingenberg

Abstract

We consider a multi-species gas mixture described by a kinetic model. More precisely, we are interested in models with Bhatnagar–Gross–Krook (BGK) interaction operators. Several extensions to the standard BGK model are studied.

Firstly, we allow the collision frequency to vary not only in time and space but also with the microscopic velocity. In the standard BGK model, the dependence on the microscopic velocity is neglected for reasons of simplicity. We allow for a more physical description by reintroducing this dependence. But even though the structure of the equations remains the same, the so-called target functions in the relaxation term become more sophisticated being defined by a variational procedure.

Secondly, we include quantum effects (for constant collision frequencies). This approach influences again the resulting target functions in the relaxation term depending on the respective type of quantum particles.

In this thesis, we present a numerical method for simulating such models. We use implicit-explicit (IMEX) time discretizations in order to take care of the stiff relaxation part due to possibly large collision frequencies. The key new ingredient is an implicit solver which minimizes a certain potential function. This procedure mimics the theoretical derivation in the models. We prove that theoretical properties of the model are preserved at the discrete level such as conservation of mass, total momentum and total energy, positivity of distribution functions and a proper entropy behavior. We provide an array of numerical tests illustrating the numerical scheme as well as its usefulness and effectiveness.

Zusammenfassung

Wir betrachten ein Gasgemisch, das aus mehreren Spezies zusammengesetzt ist und durch kinetische Modelle beschrieben werden kann. Dabei interessieren wir uns vor allem für Modelle mit BGK-Wechselwirkungsoperatoren. Verschiedene Erweiterungen des Standard-BGK-Modells werden untersucht.

Im ersten Modell nehmen wir eine Abhängigkeit der Stoßfrequenzen von der mikroskopischen Geschwindigkeit hinzu. Im Standard-BGK-Modell wird diese Abhängigkeit aus Gründen der Komplexität vernachlässigt. Wir nähern uns der physikalischen Realität weiter an, indem wir die Abhängigkeit von der mikroskopischen Geschwindigkeit beachten. Die Struktur der Gleichungen bleibt erhalten, allerdings hat dies Auswirkungen auf die sogenannten Zielfunktionen im Relaxationsterm, welche sodann durch einen Variationsansatz definiert werden.

Das zweite Modell berücksichtigt Quanteneffekte (für konstante Stoßfrequenzen), was wiederum die Zielfunktionen im Relaxationsterm beeinflusst. Diese unterscheiden sich abhängig von den jeweils betrachteten, quantenmechanischen Teilchentypen.

In dieser Doktorarbeit stellen wir numerische Verfahren vor, die auf oben beschriebene Modelle angewandt werden können. Wir legen eine implizite-explizite Zeitdiskretisierung zu Grunde, da die Relaxationsterme für große Stoßfrequenzen steif werden können. Das Kernstück ist ein impliziter Löser, der eine gewisse Potenzialfunktion minimiert. Dieses Vorgehen imitiert die theoretische Herleitung in den Modellen. Wir zeigen, dass die Eigenschaften des Modells auch auf der diskreten Ebene vorliegen. Dies beinhaltet die Massen-, Gesamtimpuls- und Gesamtenergieerhaltung, die Positivität von Verteilungsfunktionen sowie das gewünschte Verhalten der Entropie. Wir führen mehrere numerische Tests durch, die die Eigenschaften, die Nützlichkeit und die Zweckmäßigkeit des numerischen Verfahrens aufzeigen.

Contents

Acknowledgments **v**

Preface **vii**

Abstract **ix**

Contents **xi**

1 Introduction **1**

2 Theory of kinetic models **5**
2.1 Fundamentals 5
2.1.1 Models in different regimes 5
2.1.2 Transport equation 20
2.1.3 Boltzmann equation(s) 24
2.1.4 BGK equation(s) and their extensions 32
2.2 Multi-species BGK models with $\mathbf{v}$-dependent collision frequency 43
2.2.1 Motivation 43
2.2.2 A consistent BGK model with velocity-dependent collision frequency for gas mixtures 45
2.2.3 Cross sections and collision frequencies 53
2.3 Quantum multi-species BGK models 61
2.3.1 Fundamentals 61
2.3.2 A consistent multi-species quantum BGK model 66

3 Numerical schemes **81**
3.1 Fundamentals 81
3.1.1 Time discretization 81
3.1.2 Space discretization 90
3.1.3 Velocity discretization 94
3.1.4 Existing schemes for BGK equations 99
3.2 Basic idea of our scheme 101
3.3 Time discretization 105
3.3.1 First-order splitting 106
3.3.2 First-order IMEX Runge-Kutta 106
3.3.3 Second-order Strang splitting 107
3.3.4 Second-order IMEX Runge-Kutta 108
3.3.5 General implicit solver 109
3.4 Space discretization 112
3.4.1 Numerical fluxes 113

3.5 Velocity discretization 114
3.5.1 Velocity grid 114
3.5.2 Quadrature and discrete moments 114
3.5.3 Optimization algorithm 115

4 Properties of the numerical scheme **117**
4.1 Order and stability 117
4.2 Positivity of distribution functions 117
4.3 Time step restrictions 122
4.4 Conservation of mass, total momentum and total energy 123
4.5 Entropy inequality 126

5 Numerical Results **131**
5.1 Proof of order 131
5.1.1 Transport equation 131
5.1.2 Relaxation test case 133
5.2 Num. results for the classic multi-species BGK model with const. coll. frequencies 134
5.2.1 Decay rates 134
5.2.2 Hydrogen-Carbon test case 1 134
5.2.3 Sulfur-Fluorine-Electrons test case 136
5.3 Num. results for the classic multi-species BGK model with $\mathbf{v}$-dep. coll. frequencies 139
5.3.1 Illustrative toy problem 139
5.3.2 Hydrogen-Carbon test case 2 140
5.3.3 Sod problem 144
5.3.4 Mach 1.7 Shock wave problem 147
5.3.5 Mach 4 Shock wave problem 149
5.3.6 Interpenetration problem: high density 150
5.3.7 Interpenetration problem: low density 152
5.4 Numerical results for the quantum multi-species BGK model 154
5.4.1 Quantum decay rates 155
5.4.2 Sulfur-Fluorine-electrons quantum test case 157

6 Conclusions and Outlook **161**

Appendices **165**
A On damping of the electric field 165
A.1 Numerical fluxes 166
A.2 Determination of the force 167
A.3 Time step restrictions 167
A.4 Numerical tests 168
B General implicit solver for 3 species 173
C Pseudo codes 175
D Units 178
D.1 CGS system 178
D.2 Units in the quantum case 179

E List of symbols 180
F Glossary of abbreviations 183

Bibliography **185**

Chapter 1

Introduction

Mathematical fluid mechanics is a broad field describing the behavior of fluids based on partial differential equations (PDEs). 'Fluid' refers to any state of matter but solids, i.e. liquids, gases and plasmas. This includes the atmosphere, oceans, inner parts of stars and the sun. The solutions of PDEs express the behavior of such static and dynamical systems depending on provided initial and boundary conditions. This is not limited to physical phenomena, but corresponding models can also be developed for economics, finance or engineering problems. Hence, theory of PDEs is a modern field and strongly inspired by applications.

Once a model is established, we are interested in solutions of the problem. However, in general the equations cannot be solved analytically, and numerical schemes come into play. Therefore, we approximate the solutions numerically and explore properties and phenomena of the equations by numerical tests. Nevertheless, it is a difficult task to find numerical schemes which preserve the properties of the model at the discrete level. Again, this is an active field in research with relevance for science and industry.

The governing equations depend on the underlying scales of the system. Famous representatives for macroscopic scales are the Euler and Navier-Stokes equations. Whereas for rarefied gases or plasmas, which interact at mesoscopic scales, it is better to use *kinetic* models. Kinetic theory describes phenomena in statistical physics and was initialized 1872 by the physicist Ludwig Boltzmann. He developed a PDE which models the time evolution of a rarefied monatomic, single-species gas [Bol09]. The theory has been further developed among others by James Clerk Maxwell, Sydney Chapman, Thomas George Cowling, Carlo Cercignani and Cédric Villani. Last year's Nobel prize in physics (2021) was awarded to Syukuro Manabe, Klaus Hasselmann and Giorgio Parisi who have contributed to the understanding of complex systems in statistical physics [Deu20]. This illustrates the ongoing interest in this field.

In 1954, the three mathematicians Bathnagar, Gross and Krook presented a simplified model maintaining the same fundamental properties as the Boltzmann equation [BGK54]. Moreover, the model allows for efficient numerical simulations such that these equations and their extensions are of real interest for practical use.

There are several applications for kinetic theory. We mention here models for traffic flows which nowadays also explore the effect of autonomous cars [HPV22, PT21, HPRV20].

Another important application is the physical regime of plasma, and especially controlled thermonuclear fusion. The energy output from the sun originates from fusion reactions. If we succeded to rebuild a controlled 'small sun' on earth, it could support future electricity supply. One branch in fusion energy research corresponds to inertial confinement fusion (ICF). Nuclear fusion is achieved by compressing and heating targets filled with thermonuclear fuel with the help of laser beams. Only recently, a new milestone in experiments (a burning plasma) was achieved [ZHCe22]. Another branch is given by so-called Tokamak reactors in which the fusion plasma is confined by superposed magnetic fields. The energy gain and resulting power could recently be more than doubled compared to former records in the Joint European Torus (JET) experiment [Son22]. A further exciting and fascinating project in this branch is the International Thermonuclear Experimental Reactor (ITER) [Cla20]. From an economic point of view, an even larger machine than ITER will be needed. But firstly, open questions shall be answered by this project. Since fusion experiments are very challenging, numerical tests play an important role for both approaches towards controlled thermonuclear fusion.

The thesis is structured as follows. In Chapter 2, we introduce fluid models. Starting with an overview over models in several regimes in Section 2.1, we explain established kinetic models in more detail in the following sections, such as the transport equation, the Boltzmann equation and BGK equations. In Section 2.2, we consider a consistent multi-species BGK model with velocity-dependent collision frequency. We provide a motivation and a short overview over cross sections and collision frequencies. Including quantum effects opens a new field in kinetic theory. We deal with this in Section 2.3.

Chapter 3 presents numerical schemes. After a short introduction into existing schemes in Section 3.1, we give the basic idea of our scheme in Section 3.2. We carry out the details for the discretization in time in Section 3.3, for the discretization in space in Section 3.4 and for the discretization of the velocity space in Section 3.5. Our scheme is formulated in such a way that it can be applied to both the multi-species BGK model with velocity-dependent collision frequency and the quantum multi-species BGK model. Our method's core is a general implicit solver for the target functions.

In Chapter 4, we perform analytical calculations regarding properties of our numerical scheme. We show positivity of distribution functions (Section 4.2), conservation of mass, total momentum and total energy (Section 4.4) and an adequate behavior of the entropy (Section 4.5).

We present numerical results in Chapter 5. We illustrate the properties of our scheme for multi-species BGK equations with both constant and velocity-dependent collision frequencies, as well as for the quantum model. Further, we consider more physical set-ups exploring gas mixtures being relavant in plasma physics.

In Chapter 6, we provide conclusions and an outlook.

We attach several appendices. In Appendix A, we provide a short excursion which considers the coupling to a self-consistent field when charged particles are encountered. In Appendix B, we present the general implicit solver from Section 3.3.5 for 3 species. In Appendix C, we provide pseudo codes of our numerical scheme.

In Appendix D, we consider units and the unit system cm-g-s (CGS). We give a list of symbols in Appendix E and conclude with a glossary of abbreviations in Appendix F.

Chapter 2

Theory of kinetic models

Fluids are surrounding us and we make use of them in most parts of our lives. Accordingly, research in this area is of real interest. Mathematics enters in this field by establishing models as physical as possible (and by developing numerical schemes for these models which is addressed in Chapter 3). One distinguishes between microscopic, mesoscopic/kinetic and macroscopic models describing the evolution of fluids at the corresponding scales. We are interested in kinetic models where our focus lies on BGK-type models.

In Section 2.1, we give an introduction into the description of fluid models. In particular, we explain the transport equation, the Boltzmann equation and the BGK model. We present a consistent multi-species BGK model with velocity-dependent collision frequency in Section 2.2. And we conclude with a multi-species BGK model which takes quantum effects into account in Section 2.3.

2.1 Fundamentals

Starting with a short overview over existing models in different regimes in Section 2.1.1, we present selected kinetic models in more detail in Sections 2.1.2–2.1.4.

2.1.1 Models in different regimes

There exist many models for fluids. Even though these models always aim to describe the same class of matter — fluids — and are often given by (a system of) partial differential equations, they can look very different. This is not surprising because we also expect a very distinct behavior, e.g. if we think of liquids or gases. What makes the difference between those?

A basic classification comes by the mean free path and Knudsen number, respectively.

Definition 2.1.1 (Mean free path and Knudsen number). *The mean free path is an average distance which a particle travels until it collides with another particle.*

The Knudsen number Kn *is the dimensionless ratio of the mean free path and a characteristic length of the system.*

Let us consider a fluid. For a large Knudsen number ($\mathrm{Kn} \gtrsim 0.05$ [Str05]), we are able to follow each single particle and describe its *microscopic* behavior by Newton's mechanics, see Figure 2.1a.

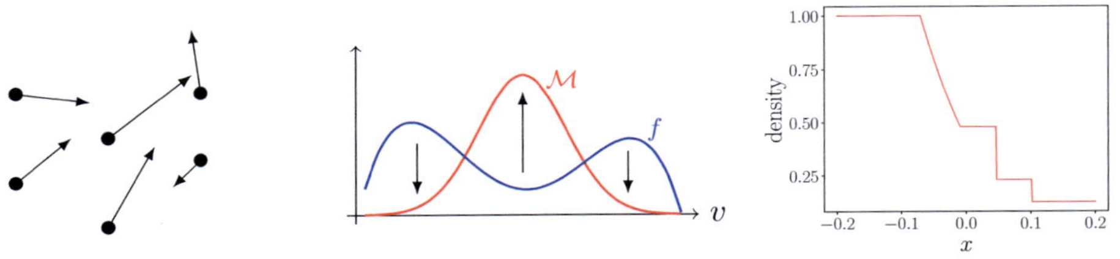

(a) microscopic description by Newton law's

(b) mesoscopic/kinetic description by distribution functions

(c) macroscopic description by macroscopic quantities

Figure 2.1: For the microscopic description of a fluid (a), each particle's position and velocity are followed. Whereas distribution functions in phase space (see Definition 2.1.12) are the basic quantity in the kinetic description (b). This plot is also explained in Section 2.1.4. In the macroscopic context (c), the number density is one quantity of interest (see Section 5.3.3 for more information about the plot).

Whereas for a very small Knudsen number ($\mathrm{Kn} \ll 1$ [Str05]), we are practically not able to follow each single particle anymore. Instead we are looking for *macroscopic* quantities which we can follow and describe. These can be measurable quantities like density, mean velocity, pressure, temperature and so on, which are statistic averages over the microscopic properties of all the particles, see Figure 2.1c. In a hydrodynamic regime, this is a valid approach in order to find a suitable and manageable description. Nevertheless, we lose information about the detailed behavior at the particle level.

What if it is not convenient to follow each particle individually, but the loss of information by the hydrodynamic approach is too severe, too? Then we are in a *mesoscopic* or *kinetic* regime. Here, we are interested in distribution functions which can be interpreted as probability distributions where the particles are located, see Figure 2.1b. So, we neither follow each single particle nor look at the statistic averages. Rather we take a description in between.

A nice summary of the description at different levels and their connection can be found in [Gra49]. In the following, we give a brief introduction to the different regimes.

Microscopic regime

In [New87], Isaac Newton formulated three axioms (in Latin language) which form the basis for the entire classical physics:

Axiom 2.1.2 (Newton's first law, principle of inertia). *'Corpus omne perseverare in statu suo quiescendi vel movendi uniformiter in directum, nisi quatenus illud a viribus impressis cogitur statum suum mutare.'*
An object remains in motion at constant speed and in a straight line unless acted on by an unbalanced force.

Axiom 2.1.3 (Newton's second law). *'Mutationem motus proportionalem esse vi motrici impressae, et fieri secundum lineam rectam qua vis illa imprimitur.'*

The time change of momentum of a body is proportional to and occurs in the same direction as the applied force.

Axiom 2.1.4 (Newton's third law, action-reaction law)**.** *'Actioni contrariam semper et aequalem esse reactionem: sive corporum duorum actiones in se mutuo semper esse aequales et in partes contrarias dirigi.'*
If an object exerts a force on another object, the second object exerts also a force on the first object which is equal in magnitude and opposite in direction.

Only in quantum mechanics and for relativistic scales, limitations of their validity are reached such that extended formulations are needed. In this project, we do not treat relativistic set-ups, but quantum theory is addressed in Section 2.3.1. For now, let us consider a fluid consisting of N (numbered) particles which obey Newton's laws. We are not interested in the description of the fluid viewed as a whole, but we aim to express the detailed happening for each particle. We follow the nice introduction in [Pir18] and examine the position $\mathbf{q}_i \in \mathbb{R}^3$ and the velocity $\mathbf{v}_i \in \mathbb{R}^3$ at any time $t \in \mathbb{R}_0^+$ for the particle with number $i \in \{1, \ldots, N\}$ and constant mass m_i (excluding chemical reactions). Then, the trajectories of each particle can be described as solutions of the following system of ordinary differential equations, formulating Axioms 2.1.2 and 2.1.3 in mathematical language.

Axiom 2.1.5 (Newton's equations)**.** *Let $\mathbf{q}_i^0$ and $\mathbf{v}_i^0$ be the position and velocity of a particle $i \in \{1, \ldots, N\}$ at time $t = t_0 \in \mathbb{R}_0^+$. For $t \geq t_0$, the evolution of the position $\mathbf{q}_i(t)$ and the velocity $\mathbf{v}_i(t)$ of this particle are given by the solution of*

$$\begin{aligned}
\frac{d}{dt}\mathbf{q}_i(t) &= \mathbf{v}_i(t), \\
\mathbf{q}_i(t_0) &= \mathbf{q}_i^0, \\
m_i \frac{d}{dt}\mathbf{v}_i(t) &= \sum_{j \neq i} \mathbf{F}_{i,j,\mathrm{int}}(\mathbf{q}_i(t), \mathbf{v}_i(t), \mathbf{q}_j(t), \mathbf{v}_j(t), t) + \mathbf{F}_{\mathrm{ext}}(\mathbf{q}_i(t), \mathbf{v}_i(t), t), \\
\mathbf{v}_i(t_0) &= \mathbf{v}_i^0.
\end{aligned} \tag{2.1}$$

The equations (2.1) describe that the change of position is prescribed by the velocity, and the change of velocity is dictated by the forces acting on the particle. The forces can be split into an internal force $\sum_{j \neq i} \mathbf{F}_{i,j,\mathrm{int}}$, describing the interactions between particles, and an external force $\mathbf{F}_{\mathrm{ext}}$.

Newton's laws of motion also establish fundamental properties whenever particles interact via elastic interactions, namely the conservation of momentum and energy during interactions. In the following, we consider binary interactions. But it is straight forward to generalize the statements to an arbitrary number of involved particles.

Theorem 2.1.6 (Conservation of momentum)**.** *Let $\mathbf{F}_{\mathrm{ext}} = 0$. Then it holds*

$$m_1 \mathbf{v}_1(t) + m_2 \mathbf{v}_2(t) = \mathrm{const} \tag{2.2}$$

for all $t \in I$, where $I \subseteq [t_0, \infty)$ is the interval in which particle 1 and particle 2 do not interact with any other of the $N-2$ particles.

Proof. Newton's equations 2.1 read

$$\begin{aligned}
m_1 \frac{d}{dt}\mathbf{v}_1(t) &= \mathbf{F}_{1,2,\mathrm{int}}(\mathbf{q}_1(t), \mathbf{v}_1(t), \mathbf{q}_2(t), \mathbf{v}_2(t), t),\\
m_2 \frac{d}{dt}\mathbf{v}_2(t) &= \mathbf{F}_{2,1,\mathrm{int}}(\mathbf{q}_2(t), \mathbf{v}_2(t), \mathbf{q}_1(t), \mathbf{v}_1(t), t)\\
&\overset{\text{Ax 2.1.4}}{=} -\mathbf{F}_{1,2,\mathrm{int}}(\mathbf{q}_1(t), \mathbf{v}_1(t), \mathbf{q}_2(t), \mathbf{v}_2(t), t)
\end{aligned}$$

for $t \in I$ provided that $\mathbf{F}_{\mathrm{ext}} = 0$. Adding both equations and integrating with respect to t yields the statement. □

For so-called *conservative* forces $\mathbf{F}$, there exists a scalar potential Φ such that $\mathbf{F} = -\nabla_{\mathbf{x}}\Phi$, $\mathbf{x}$ being the space variable. This means that the work $\int_\gamma \mathbf{F}\, \mathrm{d}s$ is independent of the path γ, and especially that the work vanishes for closed paths. So it is impossible to gain (or lose) energy by passing a closed path several times. Actually, the existence of such a scalar potential and the independence of the path are equivalent. The proof can be found in [DR11, Theorem 13.50]; and this results in the following conservation of energy.

Theorem 2.1.7 (Conservation of energy). *Let $\mathbf{F}_{\mathrm{ext}} = 0$. Assume that $\mathbf{F}_{1,2,\mathrm{int}}$ only depends on $\mathbf{q}_1(t) - \mathbf{q}_2(t)$ and that there exists a potential $\Phi(\mathbf{x})$ such that $\mathbf{F}_{1,2,\mathrm{int}} = -\nabla_{\mathbf{x}}\Phi$. Then it holds*

$$\frac{m_1}{2}|\mathbf{v}_1(t)|^2 + \frac{m_2}{2}|\mathbf{v}_2(t)|^2 + \Phi = \mathrm{const} \tag{2.3}$$

for all $t \in I$, where $I \subseteq [t_0, \infty)$ is the interval in which particle 1 and particle 2 do not interact with any other of the $N-2$ particles.

Proof. Under the given assumptions, Newton's equations (2.1) read

$$\begin{aligned}
m_1 \frac{d}{dt}\mathbf{v}_1(t) &= \mathbf{F}_{1,2,int}(\mathbf{q}_1(t) - \mathbf{q}_2(t)),\\
m_2 \frac{d}{dt}\mathbf{v}_2(t) &= \mathbf{F}_{2,1,int}(\mathbf{q}_2(t), \mathbf{v}_2(t), \mathbf{q}_1(t), \mathbf{v}_1(t), t) \overset{\text{Ax 2.1.4}}{=} -\mathbf{F}_{1,2,int}(\mathbf{q}_1(t) - \mathbf{q}_2(t))
\end{aligned}$$

for all $t \in I$. We multiply the first equation with $\mathbf{v}_1(t)$, the second one with $\mathbf{v}_2(t)$ and add both of them:

$$\begin{aligned}
\frac{d}{dt}\left(\frac{m_1}{2}|\mathbf{v}_1(t)|^2 + \frac{m_2}{2}|\mathbf{v}_2(t)|^2\right) &= \mathbf{F}_{1,2,int}(\mathbf{q}_1(t) - \mathbf{q}_2(t)) \cdot (\mathbf{v}_1(t) - \mathbf{v}_2(t))\\
&= -\nabla_{\mathbf{x}}\Phi(\mathbf{q}_1(t) - \mathbf{q}_2(t)) \cdot (\mathbf{v}_1(t) - \mathbf{v}_2(t))\\
&= -\frac{d}{dt}\Phi(\mathbf{q}_1(t) - \mathbf{q}_2(t)).
\end{aligned}$$

In the last step, we used the chain rule and Newton's equations (2.1). The statement follows by integration with respect to t. □

Newton's equations result in an additional conservation law given in the following theorem.

Theorem 2.1.8 (Conservation of angular momentum). *Let $\mathbf{F}_{\text{ext}} = 0$ and $\mathbf{b} \in \mathbb{R}^3$ be a fixed point in space. Assume that $\mathbf{F}_{1,2,\text{int}}$ is parallel to $\mathbf{q}_1(t) - \mathbf{b}$ and $\mathbf{F}_{2,1,\text{int}}$ is parallel to $\mathbf{q}_2(t) - \mathbf{b}$. Then it holds*

$$m_i(\mathbf{q}_i(t) - \mathbf{b}) \times \mathbf{v}_i(t) = \mathbf{c} = \text{const} \in \mathbb{R}^3$$

for $i = 1, 2$ and for all $t \in I$, where $I \subseteq [t_0, \infty)$ is the interval in which particle 1 and particle 2 do not interact with any other of the $N - 2$ particles. Additionally, the map $\mathbf{b} \mapsto \mathbf{c}(\mathbf{b})$ is continuous in $\mathbf{b}$.

Proof. Under the given assumptions, the product rule and Newton's equations (2.1) yield

$$\begin{aligned}\frac{d}{dt}[m_i(\mathbf{q}_i(t) - \mathbf{b}) \times \mathbf{v}_i(t)]\mathbf{v}_1(t) &= m_i(\mathbf{q}_i(t) - \mathbf{b}) \times \frac{d}{dt}\mathbf{v}_i(t) \\ &= (\mathbf{q}_i(t) - \mathbf{b}) \times \mathbf{F}_{i,j,\text{int}}(\mathbf{q}_i(t), \mathbf{v}_i(t), \mathbf{q}_j(t), \mathbf{v}_j(t), t) \\ &= 0\end{aligned}$$

for $i, j = 1, 2$, $i \neq j$ and for all $t \in I$. The first statement follows by integration with respect to t. For the continuity, let $\mathbf{b}_1, \mathbf{b}_2 \in \mathbb{R}^2$ be arbitrary. It follows

$$\begin{aligned}|\mathbf{c}(\mathbf{b}_1) - \mathbf{c}(\mathbf{b}_2)| &= |m_i(\mathbf{q}_i(t) - \mathbf{b}_1) \times \mathbf{v}_i(t) - m_i(\mathbf{q}_i(t) - \mathbf{b}_2) \times \mathbf{v}_i(t)| \\ &= |m_i(\mathbf{b}_2 - \mathbf{b}_1) \times \mathbf{v}_i(t)| \leq m_i|\mathbf{v}_i(t)||\mathbf{b}_2 - \mathbf{b}_1|.\end{aligned}$$

Hence, $\mathbf{c}(\mathbf{b})$ is Lipschitz continuous in $\mathbf{b}$, in particular continuous. □

The physical consequence is that particles stay in the same plane while colliding. This becomes more obvious in the following corollary.

Corollary 2.1.9 (Post-collisional velocities). *Let $\mathbf{F}_{\text{ext}} = 0$. Assume that $\mathbf{F}_{1,2,\text{int}}$ is parallel to $\mathbf{q}_1(t) - \mathbf{q}_2(t)$, $\mathbf{F}_{2,1,\text{int}}$ is parallel to $\mathbf{q}_2(t) - \mathbf{q}_1(t)$, and let there exist a compactly supported potential $\Phi(\mathbf{x})$ such that $\mathbf{F}_{1,2,\text{int}} = -\nabla_{\mathbf{x}}\Phi$. During the interaction, we assume that only particle 1 and particle 2 are involved. Furthermore, we assume that the interaction occurs instantaneously at time $t = t_*$ when the particles reached their minimal distance. We call $\boldsymbol{\xi}$ the unit vector along the line with minimal distance of the two particles during the interaction in direction of particle 2, see Figure 2.2. Then $\boldsymbol{\xi}$ can be written as*

$$\boldsymbol{\xi} = \frac{\mathbf{q}_2(t_*) - \mathbf{q}_1(t_*)}{|\mathbf{q}_2(t_*) - \mathbf{q}_1(t_*)|}$$

and the conservation laws during the collision read

$$\begin{aligned}m_1\mathbf{v}_1 + m_2\mathbf{v}_2 &= m_1\mathbf{v}_1' + m_2\mathbf{v}_2', \\ m_1|\mathbf{v}_1|^2 + m_2|\mathbf{v}_2|^2 &= m_1|\mathbf{v}_1'|^2 + m_2|\mathbf{v}_2'|^2\end{aligned}$$

with the velocities $\mathbf{v}_1'$ *and* $\mathbf{v}_2'$ *after the interaction*

$$\mathbf{v}_1' = \mathbf{v}_1 - \frac{2m_2}{m_1 + m_2}[(\mathbf{v}_2 - \mathbf{v}_1) \cdot \boldsymbol{\xi}]\boldsymbol{\xi},$$
$$\mathbf{v}_2' = \mathbf{v}_2 + \frac{2m_1}{m_1 + m_2}[(\mathbf{v}_2 - \mathbf{v}_1) \cdot \boldsymbol{\xi}]\boldsymbol{\xi}.$$

Proof. The proof can be found e.g. in [Pir18] and follows from the previous theorems. □

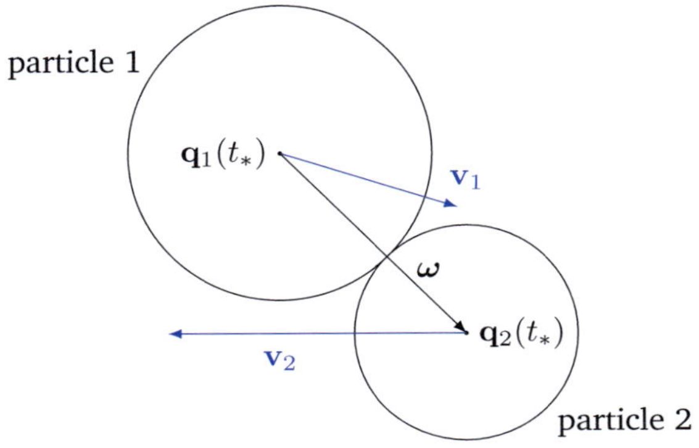

Figure 2.2: At time $t = t_*$, particle 1 and particle 2 reach their minimal distance such that they collide. The vector $\boldsymbol{\omega}$ connects the particles' positions $\mathbf{q}_1(t_*)$ and $\mathbf{q}_2(t_*)$, and $\boldsymbol{\xi} = \frac{\boldsymbol{\omega}}{|\boldsymbol{\omega}|}$.

At the microscopic level, elastic collisions are time reversible in the following sense. Consider two particles with velocities $\mathbf{v}_1$ and $\mathbf{v}_2$. Due to a collision, they exchange momentum and energy, and their velocities convert into $\mathbf{v}_1'$ and $\mathbf{v}_2'$, respectively. If we now move backward in time, change $\mathbf{v}_i \mapsto -\mathbf{v}_i'$ and $\mathbf{v}_i' \mapsto -\mathbf{v}_i$, all theorems still hold true. So there is no preferred direction for collisions to happen. This is in contrast to the descriptions at mesoscopic and macroscopic levels. We will emphasize this in Section 2.1.3.

Knowing the initial positions and velocities of all particles in a gas, we can calculate the evolution of a classic gas for all times by Newton's laws. When this is solved numerically, it is often called *molecular dynamics*. However, this is a very rich description. Only $1\,\mathrm{g}$ hydrogen already contains about $6 \cdot 10^{23}$ particles. This illustrates that the detailed description of Newton's mechanics might be useful under specific circumstances, but in many applications it is not practicable. This leads us to the kinetic regime.

Mesoscopic/kinetic regime

The description at the microscopic level is very rich and also discrete. In order to approach a continuum, we consider an ensemble of conceptual copies of a given physical system. Then we look at the probability W that the system is in the corresponding state at a given time. [Gra49]

Let N be fixed and let us assume N particles of mass m whose position is denoted by $\mathbf{q}_i$ and whose momentum is given by $\mathbf{p}_i$ for $i = 1, \dots, N$, as it is conventionally done in theoretical mechanics. The probability that the system is located in the volume $\mathrm{d}\mathbf{q}_1 \dots \mathrm{d}\mathbf{q}_N \, \mathrm{d}\mathbf{p}_1 \dots \mathrm{d}\mathbf{p}_N$ is represented by the quantity

$$W = W(\mathbf{q}_1, \dots, \mathbf{q}_N, \mathbf{p}_1, \dots, \mathbf{p}_N).$$

The following statements can be found e.g. in [Sch06, Cer88, LL69].

Theorem 2.1.10 (Liouville's theorem). *The function W is constant along the trajectories of the system. In other words, the volume and orientation of a flow in the $(\mathbf{q}, \mathbf{p})$-space stays constant in time.*

From Liouville's theorem, it follows Liouville's equation

$$\begin{aligned} \partial_t W + \sum_{i=1}^{N} \frac{\mathbf{p}_i}{m} \cdot \nabla_{\mathbf{q}_i} W + \sum_{i=1}^{N} \frac{\mathbf{F}_i}{m} \cdot \nabla_{\mathbf{p}_i} W = 0, \\ \mathbf{F}_i = -\sum_{j \neq i}^{N} \nabla_{\mathbf{q}_i} \left(\phi_{ij} + \phi_{\text{ext}} \right), \end{aligned} \tag{2.4}$$

where the force $\mathbf{F}_i$ on particle i is determined by the gradient of a potential which is the sum of an external potential ϕ_{ext} and an internal potential ϕ_{ij} coming from interactions between the particles i and j.

We integrate Liouville's equation (2.4) over parts of the variables and build a chain of equations. This is called the Bogoliubov–Born–Green–Kirkwood–Yvon (BBGKY) hierarchy [Bog46, BG46, Kir46, Kir47, Yvo35]. We consider the functions [CIP94]

$$\begin{aligned} &f_s(\mathbf{q}_1, \dots, \mathbf{q}_s, \mathbf{p}_1, \dots, \mathbf{p}_s) \\ &\qquad = \int W(\mathbf{q}_1, \dots, \mathbf{q}_N, \mathbf{p}_1, \dots, \mathbf{p}_N) \, \mathrm{d}\mathbf{q}_{s+1} \dots \mathrm{d}\mathbf{q}_N \, \mathrm{d}\mathbf{p}_{s+1} \dots \mathrm{d}\mathbf{p}_N \end{aligned}$$

for $s = 1, \dots, N$. The probable number of particles in e.g. the volume $\mathrm{d}\mathbf{q}_1 \, \mathrm{d}\mathbf{p}_1$ is given by

$$N f_1 \mathrm{d}\mathbf{q}_1 \, \mathrm{d}\mathbf{p}_1.$$

We obtain N equations where the time evolution of f_s is given by

$$\begin{aligned} \partial_t f_s + \sum_{i=1}^{s} \frac{\mathbf{p}_i}{m} \cdot \nabla_{\mathbf{q}_i} f_s + \sum_{i=1}^{s} \frac{\mathbf{F}_i}{m} \cdot \nabla_{\mathbf{p}_i} f_s \\ = (N - s) \sum_{i=1}^{s} \int (\nabla_{\mathbf{q}_i} \phi_{i,s+1}) \cdot (\nabla_{\mathbf{p}_i} f_{s+1}) \, \mathrm{d}q_{s+1} \, \mathrm{d}p_{s+1} \\ \mathbf{F}_i = -\sum_{j \neq i}^{s} \nabla_{\mathbf{q}_i} \left(\phi_{ij} + \phi_{\text{ext}} \right) \end{aligned}$$

with a correction term on the right-hand side (RHS) representing the influence of the $N - s$ suppressed particles. In this chain, the s-th equation connects f_s and f_{s+1}. So, in order to determine only f_1 one has to solve the equation for f_2, which requires the solution of f_3, and so on, and eventually one needs to solve the full Liouville equation.

However, we can break the chain at some point and solve the equation for f_s if we found a sufficient approximation for f_{s+1}. For instance, after appropriate assumptions the marginal f_1 becomes a kinetic distribution function (see Definition 2.1.12), and an adequate estimate of $f_2(\mathbf{q}_1, \mathbf{q}_2, \mathbf{p}_1, \mathbf{p}_2)$ turns into the collision operator of the Boltzmann equation (2.25) [Gra49, CIP94]. Here, the Boltzmann-Grad limit enforces

$$N \to \infty \quad \text{and} \quad r \to 0 \quad \text{while} \quad Nr^2 \to \text{const} \in \mathbb{R}^+.$$

We do not discuss the limiting process in this thesis. But in Section 2.1.3, we give a motivation where the famous Boltzmann equation comes from.

Remark 2.1.11 (Irreversibility). *While the microscopic description is time reversible, the mesodynamics (and macrodynamics) behave irreversible. The irreversibility enters the description when the BBGKY chain gets broken and interpreted stochastically [Cer88]. This issue led to discussions, especially concerning the Boltzmann equation, see Section 2.1.3.*

The previously given equations are supposed to motivate the description at the mesoscopic level. It is based on a function which combines all particles. Since we do not want to track each particle, we look for a dependence only on space $\mathbf{x}$, microscopic velocity $\mathbf{v}$ and time t (not on each particle's position and velocity). In the following, be aware that motivations and discussions using individual particles only serve for illustration purposes. Kinetic theory does not consider particles themselves, but the so-called distribution function.

Definition 2.1.12 (Distribution function and phase space). *A function $f : \mathbb{R}^3 \times \mathbb{R}^3 \times \mathbb{R}_0^+ \to \mathbb{R}_0^+$ is called a distribution function if and only if $f(\mathbf{x}, \mathbf{v}, t)\,\mathrm{d}\mathbf{x}\,\mathrm{d}\mathbf{v}$ is the number of particles with velocities in $(\mathbf{v}, \mathbf{v} + \mathrm{d}\mathbf{v})$ located at $(\mathbf{x}, \mathbf{x} + \mathrm{d}\mathbf{x})$ at time t.*

We call $(\mathbf{x}, \mathbf{v})$ the phase space.

The value of the distribution function $f(\mathbf{x}, \mathbf{v}, t)$ can also be interpreted as the probability with which a particle with velocity $\mathbf{v}$ can be found at position $\mathbf{x}$ and time t.

Definition 2.1.13 (General kinetic equation). *A general kinetic equation can be written as*

$$\partial_t f + \mathbf{v} \cdot \nabla_{\mathbf{x}} f + \frac{\mathbf{F}(\mathbf{x}, t)}{m} \cdot \nabla_{\mathbf{v}} f = \mathcal{Q}[f]. \tag{2.5}$$

The left-hand side (LHS) of (2.5) describes that particles move uniformly along straight lines, but they might be deflected by a force $\mathbf{F}$. We consider this part of the equation in Section 2.1.2. The RHS of (2.5) takes care of possible interactions between the particles. This general formulation captures many different models. In the following, we list possible variations for this equation.

For stationary problems, it holds $\partial_t f = 0$. The second term on the LHS vanishes in space homogeneous settings. And often it is $\mathbf{F} = 0$, otherwise the force $\mathbf{F}$ is either given or needs to be determined separately.

Moreover, the RHS takes many different forms focusing on diverse aspects of possible interactions between particles. We will give a short overview over kinetic models in the following and illustrate it in Figure 2.3.

If direct interactions are neglected (being valid for short time scales), we have $Q[f] = 0$ which is the case e.g. for the Vlasov equation. This is often used for plasmas which means that charged particles are encountered and the force term due to a self-consistent field becomes crucial. For special geometries and applications as the Tokamak, the gyrokinetic coordinates have been developed [GIVW10].

The specific treatment of interactions becomes necessary for long periods [Vil02] and leads to different collision operators on the RHS of (2.5) and to different models, respectively. From the BBGKY hierarchy, the Boltzmann equation can be derived by a binary scattering approximation. This model is discussed in Section 2.1.3.

Another approach is a correlation expansion [Len60, Bal60]. This yields the Lenard-Balescu equation with the interaction operator

$$\mathcal{Q}[f] = \nabla_{\mathbf{v}} \cdot \int K(\mathbf{v}, |\mathbf{v} - \mathbf{v}_*|, \nabla_{\mathbf{v}} f) \left[f(\mathbf{v}_*) \nabla_{\mathbf{v}} f(\mathbf{v}) - (\nabla_{\mathbf{v}_*} f(\mathbf{v}_*)) f(\mathbf{v}) \right] \mathrm{d}\mathbf{v}_*$$

and a strong nonlocal and nonlinear kernel $K(\mathbf{v}, |\mathbf{v} - \mathbf{v}_*|, \nabla_{\mathbf{v}} f)$. The Lenard-Balescu equation intrinsically includes screening given by a dynamical dielectric response [PB52]. This is why, in plasma physics, it is often considered to be more suitable than the Boltzmann equation (where modifications need to be done in order to handle charged particles, see Remark 2.2.9). However, the nonlinear kernel makes the model very complicated. For more details, see [Str06] and references therein.

Both the Boltzmann and the Lenard-Balescu approach can be continued to the Landau equation when grazing collisions dominate [Lan37, RMJ57, Len60]. It incorporates the interaction operator

$$\mathcal{Q}[f] = \nabla_{\mathbf{v}} \cdot \int K(|\mathbf{v} - \mathbf{v}_*|) \left[f(\mathbf{v}_*) \nabla_{\mathbf{v}} f(\mathbf{v}) - (\nabla_{\mathbf{v}_*} f(\mathbf{v}_*)) f(\mathbf{v}) \right] \mathrm{d}\mathbf{v}_*.$$

Compared to the Lenard-Balescu equation, the kernel $K(|\mathbf{v} - \mathbf{v}_*|)$ is much simpler and better handable. It contains the so-called Coulomb logarithm which indicates the preponderance of many weak interactions, cf. Remark 2.2.10. The asymptotic limit for the derivation of the Landau equation has been justified among others in [DLD92, Des92]. In special cases and coordinate systems, the nonlinear Landau equation reduces to the linear Fokker-Planck equation [Vil02], and the names are often coupled to Fokker-Planck/Landau equation. In plasma physics, the Fokker-Planck/Landau equation is often used [LM05].

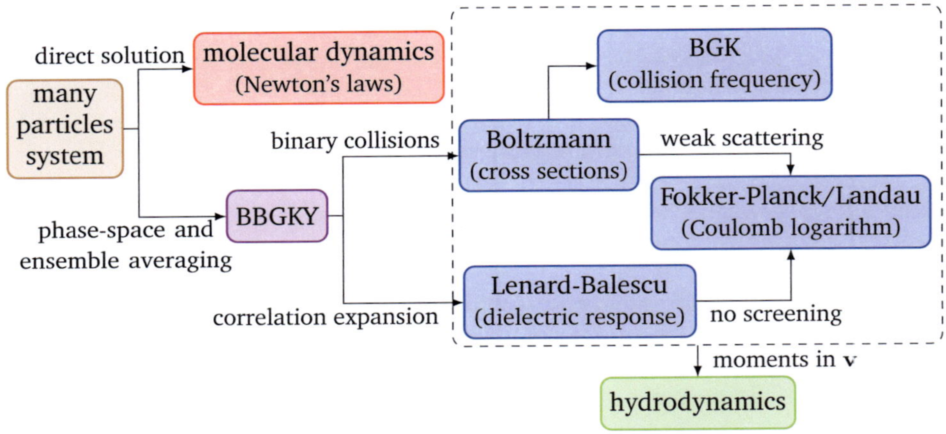

Figure 2.3: An overview over the structure in kinetic theory is given, inspired by [SM16]. Instead of solving Newton's equations (2.1) directly, the system of equations can be approximated by the BBGKY hierarchy. If the number of particles goes to infinity while conserving mass, kinetic models can formally be derived. A correlation expansion yields the Lenard-Balescu model representing many-body physics in the weak-scattering limit established by a dielectric response. Alternatively, a binary-collision assumption leads to the Boltzmann equation where the detailed physics is hidden in the cross sections. Both branches can be continued to the Fokker-Planck/Landau equation for weak- and binary-scattering collisions where the Coulomb logarithm appears as important quantity. We are especially interested in BGK models which can be derived from the Boltzmann equation where the cross sections convert to collision frequencies. A hydrodynamic description can be obtained by taking moments in $\mathbf{v}$ of the kinetic equations.

The kinetic formulation is useful for the mesoscopic regime. However, the distribution function f is a mathematical construction and not physically measurable. The distribution function is yet linked to measurable quantities given by moments of f.

Definition 2.1.14 (Macroscopic quantities). *Let S be a set of indices, each index corresponding to a species. Let $f_i : \mathbb{R}^3 \times \mathbb{R}^3 \times \mathbb{R}_0^+ \to \mathbb{R}_0^+$ with $(1+|v|^2)f_i \in L^1(\mathrm{d}\mathbf{v})$ be the distribution function of species $i \in S$.*

1. *The functions*

$$n_i : \mathbb{R}^3 \times \mathbb{R}_0^+ \to \mathbb{R}_0^+, \quad (\mathbf{x}, t) \mapsto \int_{\mathbb{R}^3} f_i(\mathbf{x}, \mathbf{v}, t)\,\mathrm{d}\mathbf{v} \quad \text{and}$$
$$\rho_i : \mathbb{R}^3 \times \mathbb{R}_0^+ \to \mathbb{R}_0^+, \quad (\mathbf{x}, t) \mapsto m_i \int_{\mathbb{R}^3} f_i(\mathbf{x}, \mathbf{v}, t)\,\mathrm{d}\mathbf{v}$$

   are called number density and mass density of species i, respectively.

2. *We define*

$$n_i\mathbf{u}_i : \mathbb{R}^3 \times \mathbb{R}_0^+ \to \mathbb{R}, \quad (\mathbf{x}, t) \mapsto \int_{\mathbb{R}^3} f_i(\mathbf{x}, \mathbf{v}, t)\mathbf{v}\,\mathrm{d}\mathbf{v}.$$

   For $n_i > 0$, the function $\mathbf{u}_i = \frac{n_i\mathbf{u}_i}{n_i}$ is called the mean velocity of species i.

3. *The function*

$$E_i : \mathbb{R}^3 \times \mathbb{R}_0^+ \to \mathbb{R}_0^+, \quad (\mathbf{x}, t) \mapsto \frac{m_i}{2} \int_{\mathbb{R}^3} f_i(\mathbf{x}, \mathbf{v}, t) |\mathbf{v}|^2 \, \mathrm{d}\mathbf{v}$$

is said to be the energy density of species i.

4. *The internal energy of species i is given by*

$$e_i : \mathbb{R}^3 \times \mathbb{R}_0^+ \to \mathbb{R}_0^+,$$
$$(\mathbf{x}, t) \mapsto E_i(\mathbf{x}, t) - \frac{m_i}{2} n_i |\mathbf{u}_i|^2 = \frac{m_i}{2} \int_{\mathbb{R}^3} f_i(\mathbf{x}, \mathbf{v}, t) |\mathbf{v} - \mathbf{u}_i|^2 \, \mathrm{d}\mathbf{v}.$$

5. *For an ideal gas and $n_i > 0$, the (kinetic) temperature*[1] *of species i is defined by*

$$T_i : \mathbb{R}^3 \times \mathbb{R}_0^+ \to \mathbb{R}_0^+, \quad (\mathbf{x}, t) \mapsto \frac{2e_i}{dn_i} = \frac{m_i}{dn_i} \int_{\mathbb{R}^3} f_i(\mathbf{x}, \mathbf{v}, t) |\mathbf{v} - \mathbf{u}_i|^2 \, \mathrm{d}\mathbf{v} \qquad (2.6)$$

where d denotes the dimensions in velocity space. In this thesis, it is $d = 3$.

6. *The energy flux of species i is defined to be*

$$\mathbf{Q}_i : \mathbb{R}^3 \times \mathbb{R}_0^+ \to \mathbb{R}^3, \quad (\mathbf{x}, t) \mapsto \frac{m_i}{2} \int_{\mathbb{R}^3} f_i(\mathbf{x}, \mathbf{v}, t) |\mathbf{v}|^2 \mathbf{v} \, \mathrm{d}\mathbf{v}.$$

7. *We call*

$$\mathbb{P}_i : \mathbb{R}^3 \times \mathbb{R}_0^+ \to \mathbb{R}^{3\times 3}, \quad (\mathbf{x}, t) \mapsto m_i \int_{\mathbb{R}^3} f_i(\mathbf{x}, \mathbf{v}, t) (\mathbf{v} - \mathbf{u}_i) \otimes (\mathbf{v} - \mathbf{u}_i) \, \mathrm{d}\mathbf{v}$$

the pressure tensor of species i. The pressure of species i is given by

$$p_i = \frac{1}{3} \operatorname{tr} \mathbb{P}_i.$$

Remark 2.1.15 (Degrees of freedom). *The formula for the temperature* (2.6) *depends on the translational degrees of freedom which comes by the dimensionality of the* $\mathbf{v}$*-space. We consider three independent* $\mathbf{v}$*-variables corresponding to three translational degrees of freedom, $d = 3$. Since we assume a monatomic gas, there are no rotational degrees of freedom. In more general cases, the notion of temperature needs to be extended.*

If more than one species is involved, macroscopic quantities of the gas mixture need to be defined. Let S be the set of indices for the species. For N species, it is $S = \{1, \ldots, N\}$. Be aware that N is not the number of particles but the number of different species.

[1] We suppress the Boltzmann constant k_B for ease of presentation. If the Boltzmann constant is required, the replacement $T \mapsto k_B T$ works in most cases which can be verified by a check of units. In Section 2.2.3, we emphasize the use of k_B for clearness.

Definition 2.1.16 (Mixture quantities). *The total number density and total mass density are given by*

$$n_{\text{mix}} = \sum_{i \in S} n_i \quad \textit{and} \quad \rho_{\text{mix}} = \sum_{i \in S} \rho_i. \tag{2.7}$$

For nonvanishing ρ_{mix}, we denote the mixture mean velocity

$$\mathbf{u}_{\text{mix}}(\mathbf{x}, t) = \frac{\sum_{i \in S} \rho_i \mathbf{u}_i}{\rho_{\text{mix}}} \tag{2.8}$$

and the total energy

$$E_{\text{mix}} = \frac{3}{2} n_{\text{mix}} T_{\text{mix}} + \frac{1}{2} \rho_{\text{mix}} |\mathbf{u}_{\text{mix}}|^2$$

where the mixture temperature reads

$$\begin{aligned} T_{\text{mix}}(\mathbf{x}, t) &= \frac{2}{3 n_{\text{mix}}} \sum_{i \in S} \int \frac{m_i}{2} |\mathbf{v} - \mathbf{u}_{\text{mix}}|^2 f_i \,\mathrm{d}\mathbf{v} \\ &= \frac{1}{n_{\text{mix}}} \left(\sum_{i \in S} n_i T_i + \frac{1}{3} \sum_{i \in S} \rho_i (|\mathbf{u}_i|^2 - |\mathbf{u}_{\text{mix}}|^2) \right) \\ &= \frac{1}{n_{\text{mix}}} \left(\sum_{i \in S} n_i T_i + \frac{1}{3 \rho_{\text{mix}}} \sum_{\substack{i,j \in S \\ i<j}} \rho_i \rho_j (|\mathbf{u}_i - \mathbf{u}_j|^2) \right). \end{aligned} \tag{2.9}$$

Moreover, the reduced mass is defined by

$$\mu_{ij} = \frac{m_i m_j}{m_i + m_j}. \tag{2.10}$$

Taking moments of f means that we average over the velocities. This leads to a loss of the detailed information in f. Nevertheless, in many occasions the description by these averaged values is sufficient.

Macroscopic regime

The macroscopic regime seems to be the most natural one because we can feel and see the corresponding quantities defined in Definition 2.1.14. That is why we are also familiar with (physical) measurements at this scale in our daily life. Nevertheless, as we have seen in the previous sections, this is only one possible description.

We can formally derive hydrodynamic equations from the kinetic description even though rigorous proofs have only been found for specific limits [SR09]. One possibility for the formal derivation are moment methods, another approach is the Chapman-Enskog expansion [Str05] developped independently by Chapman [Cha16, Cha18] and Enskog [Ens17, Ens21]. Starting e.g. from the Boltzmann equation, setting the distribution function to be Maxwellian (in equilibrium) and taking moments, we get the well-known Euler equations. For more details, we refer to [Str05, SR09].

Definition 2.1.17 (Euler equations). *Let $\rho, \mathbf{u}, e$ be a fluid's mass density, mean velocity and internal energy and sufficiently smooth. Let the pressure p be a function of the internal energy. Then the Euler equations read*

$$\begin{aligned} \partial_t \rho + \nabla_{\mathbf{x}} \cdot (\rho \mathbf{u}) &= 0, \\ \partial_t (\rho \mathbf{u}) + \nabla_{\mathbf{x}} \cdot (\rho \mathbf{u} \otimes \mathbf{u} + p \mathbb{1}) &= 0, \\ \partial_t \left(\frac{\rho}{2} |\mathbf{u}|^2 + e \right) + \nabla_{\mathbf{x}} \cdot \left(\frac{\rho}{2} |\mathbf{u}|^2 + e + p \right) \mathbf{u} &= 0. \end{aligned} \tag{2.11}$$

We do not specify the required smoothness of the solutions here because there exist several frameworks such as *strong* solutions, *weak* solutions and *measure-valued* solutions. Every notion lowers the required regularity, see [Mar21] and references therein.

The Euler equations are one of the oldest systems of partial differential equations describing the motion of fluids, formulated by Euler in the 18th century [Eul57]. They also correspond to the zeroth order in the Chapman-Enskog expansion. Considering first-order corrections, we obtain the famous Navier-Stokes equations. These include viscosity and heat conductivity, so they capture more phenomena in fluids.

Definition 2.1.18 (Navier-Stokes equations). *Let $\rho, \mathbf{u}, p$ be a fluid's mass density, mean velocity and pressure and sufficiently smooth. Let the pressure p be a function of the internal energy. We denote the viscosity by μ and heat conductivity by κ. Then the Navier-Stokes equations read*

$$\begin{aligned} \partial_t \rho + \nabla_{\mathbf{x}} \cdot (\rho \mathbf{u}) &= 0 \\ \partial_t (\rho \mathbf{u}) + \nabla_{\mathbf{x}} \cdot (\rho \mathbf{u} \otimes \mathbf{u} + p \mathbb{1}) &= \nabla_{\mathbf{x}} \cdot \sigma \\ \partial_t \left(\frac{\rho}{2} |\mathbf{u}|^2 + e \right) + \nabla_{\mathbf{x}} \cdot \left(\frac{\rho}{2} |\mathbf{u}|^2 + e + p \right) \mathbf{u} &= \kappa \Delta_{\mathbf{x}} T + \nabla_{\mathbf{x}} \cdot (\sigma \cdot \mathbf{u}) \end{aligned} \tag{2.12}$$

with the trace-free stress tensor

$$\sigma = 2\mu \left(\nabla_{\mathbf{x}} \mathbf{u} + (\nabla_{\mathbf{x}} \mathbf{u})^\top - \frac{1}{3} (\nabla_{\mathbf{x}} \cdot \mathbf{u}) \mathbb{1} \right).$$

The viscosity and thermal conductivity belong to the so-called *transport coefficients* for macroscopic laws which can be measured by experiments. The Chapman-Enskog expansion allows to compute them from microscopic details of the gas [Str05]. A constant, which will be important in the context of the BGK equations, is the Prandtl number.

Definition 2.1.19 (Prandtl number). *The Prandtl number* Pr *is the dimensionless ratio of viscosity and heat conductivity*

$$\mathrm{Pr} = \frac{5 k_B}{2m} \frac{\mu}{\kappa}$$

where k_B denotes the Boltzmann constant and m the mass of the gas. The measured value for a monatomic gas is $\mathrm{Pr} \approx \frac{2}{3}$.

The systems of equations (2.11) and (2.12) represent conservation laws at the macroscopic level.

Theorem 2.1.20. *The Euler equations* (2.11) *and Navier-Stokes equations* (2.12) *conserve mass, momentum and energy.*

Proof. For the Euler equations, this gets evident when integrating with respect to $\mathbf{x}$ and applying Gauß's law. Then the mass $\int \rho \,\mathrm{d}\mathbf{x}$, the momentum $\int \rho \mathbf{u} \,\mathrm{d}\mathbf{x}$ and the energy $\int \left(\frac{\rho}{2}|\mathbf{u}|^2 + e\right) \,\mathrm{d}\mathbf{x}$ are conserved unless there is a flux over the boundary. For the Navier-Stokes equations, more computations are required. We refer to [TM05, LL91]. □

One requirement of the systems (2.11) and (2.12), respectively, is the so-called *equation of state* which closes the systems and relates the pressure p to the internal energy e, respective density ρ and temperature T. Thermodynamic theory enters the fluid dynamics among other things by these laws. For ideal gases, it holds

$$p = (\gamma - 1)\, e$$

with the ratio $\gamma = \frac{\mathfrak{f}+2}{\mathfrak{f}}$ and the degrees of freedom $\mathfrak{f}$. We consider three translational degrees of freedom but no rotations nor vibrations; that is $\mathfrak{f} = d = 3$ and $p = nT$. Here, we slightly abused notation because this ideal gas law directly follows from kinetic gas theory [Bol09]; we wanted to give the equations in a bigger context where also more general equations of state can be applied even if we do not go into further details.

In order to obtain as physical solutions as possible, additional properties and admissibility conditions may be required. One notion, which will also be important for our kinetic models, is entropy. In the hydrodynamic set-up, it is linked to an additional inequality to be fulfilled by the system. This specifies a preferred direction in time and excludes solutions with unphysical properties. For further information, we refer to [Eva10, Mar21].

There exist many variations for the systems (2.11) and (2.12). We already mentioned the possibility of different equations of state, but also source terms can be taken into account, e.g. gravitation in stars. One also distinguishes between compressible and incompressible equations, where for incompressible conservation laws it is $\nabla_{\mathbf{x}} \cdot \mathbf{u} = 0$.

The given systems consider fluids consisting only of one species. If more species are involved, this refers to global quantities corresponding to the assumption that every species behaves similarly. However, this approach is not valid when the species behave differently, and multi-species equations need to be considered. In this case, each species follows its equations for mass, momentum and energy. The mass of each species, total momentum and total energy are conserved. The exchange of momentum and energy between the species introduces a coupling of equations of different species. [RS07]

The Euler and Navier-Stokes equations are a field of active research. Very famous is the open question on the existence and uniqueness of solutions for the Navier-Stokes equations. This is stated to be a Millenium Prize Problem [Ins] which is why this issue is known even outside the field. However, already the Euler equations let us tumble regarding the existence and uniqueness of solutions. Recent results prove that there exist initial data for which the Euler equations (in multiple space dimensions) admit infinitely many solutions which do fulfil an entropy inequality. [Mar21]

Another topic is the development of adequate numerical schemes. For example, there are situations where it is necessary to maintain a stationary solution exactly. For this purpose, *well-balanced* schemes are developed [BCK21]. Another feature of interest is that a discretization stays accurate and efficient throughout several regimes. By this comment, we close the circle and observe that the macroscopic equations, representing the hydrodynamic limit, are also of interest for the kinetic regime. As a scaling parameter (e.g. the Knudsen number) approaches zero, we formally pass over from the kinetic to the macroscopic description. Numerical formulations shall be uneffected by this transition which is called asymptotic-preserving (AP). We will elaborate this in Chapter 3.

We end this short overview with a brief section on thermodynamics.

Thermodynamics and statistical mechanics

Thermodynamics is a branch of physics mainly dealing with the macroscopic quantities heat, work and temperature. Introduced by physical experiments and observations at macroscopic scales, it is being consolidated by statistical mechanics which describes physical properties of many-body systems on the basis of microscopic principles. This surely reminds us of the kinetic description of fluids, and it is not surprising to detect links between these theories.

Comparable to Newton's laws in microdynamics, the fundamental laws of thermodynamics form the basis of this subject. The first one concerns the conservation of energy [Sch06]. We here give the second one.

Axiom 2.1.21 (Second law of thermodynamics). *'Wärme kann nie von selbst von einem kälteren in ein wärmeres Reservoir übergehen.' (Rudolf Clausius, 1850) [Sch06] Heat can only pass from a colder to a warmer reservoir with another change connected to it.*

There exist different but equivalent formulations of the second law of thermodynamics, e.g. by Lord Kelvin in 1851. In the end, they all result in the existence of reversible and irreversible processes connected to a basic quantity in statistical physics, called entropy S. A mathematical formulation reads

$$\vartheta\frac{\mathrm{d}S}{\mathrm{d}t} \geq \frac{\mathrm{d}E}{\mathrm{d}t} + p\frac{\mathrm{d}V}{\mathrm{d}t} \tag{2.13}$$

where ϑ denotes the absolute temperature, E the energy of the system, p the pressure and V the volume occupied by the gas.

One of the entropy's main features is that it either stays constant or increases in time. If the entropy increases, we consider an irreversible process being in contrast to the overall reversible microdynamics.

For more details, we refer to the textbook [Sch06].

We will later see that entropy plays an essential role in kinetic theory. The second law of thermodynamics 2.1.21, being introduced as axiom, can be derived and formulated as theorem which is known as $\mathcal{H}$-Theorem. This was firstly done for the famous Boltzmann equation by Boltzmann himself, see Theorem 2.1.31.

Before we go into detail, we study another kinetic equation which partly appears in the Boltzmann equation as well.

2.1.2 Transport equation

The general kinetic equation (2.5) consists of two parts. The LHS describes the transport in phase space, whereas the RHS specifies the interactions of the particles. In this section, we focus on the LHS with vanishing RHS, the so-called transport equation. Many introductory books explain such kind of equation, e.g. [Joh71, Chapters 1.3 and 1.4] and [Eva10, Chapter 1.2]. We only consider the following special form of the transport equation which, in particular, is hyperbolic.

Definition 2.1.22 (Transport equation). *The homogeneous transport equation reads*

$$\begin{aligned} \partial_t f + \mathbf{v}\cdot\nabla_{\mathbf{x}} f + \frac{\mathbf{F}(\mathbf{x},t)}{m}\cdot\nabla_{\mathbf{v}} f &= 0, \\ f(\mathbf{x},\mathbf{v},0) &= f^0(\mathbf{x},\mathbf{v}). \end{aligned} \tag{2.14}$$

By the method of characteristics, the partial differential equation (2.14) can be reduced to a system of ordinary differential equations.

Definition 2.1.23 (Characteristic curve). *A curve along which the solution of* (2.14) *is constant is called characteristic curve. The corresponding system of ordinary differential equations is called the characteristic equations.*

In the following, we perform this approach for (2.14). Let

$$\gamma : [0,1] \to \mathbb{R}^3 \times \mathbb{R}^3 \times \mathbb{R}_0^+, \quad s \mapsto \gamma(s) = (\mathbf{x}(s), \mathbf{v}(s), t(s))$$

be a smooth parametrization of a curve in $\mathbb{R}^3 \times \mathbb{R}^3 \times \mathbb{R}_0^+$. For $q(s) = f(\gamma(s)) = f(\mathbf{x}(s), \mathbf{v}(s), t(s))$ it is

$$\frac{\mathrm{d}}{\mathrm{d}s}q(s) = \frac{\mathrm{d}t(s)}{\mathrm{d}s}\partial_t f + \frac{\mathrm{d}\mathbf{x}(s)}{\mathrm{d}s}\cdot\nabla_{\mathbf{x}} f + \frac{\mathrm{d}\mathbf{v}(s)}{\mathrm{d}s}\cdot\nabla_{\mathbf{v}} f.$$

If γ is a solution to

$$\begin{aligned}\frac{\mathrm{d}t(s)}{\mathrm{d}s} &= 1,\\ \frac{\mathrm{d}\mathbf{x}(s)}{\mathrm{d}s} &= \mathbf{v}(s),\\ \frac{\mathrm{d}\mathbf{v}(s)}{\mathrm{d}s} &= \frac{\mathbf{F}(\mathbf{x}(s), t(s))}{m},\end{aligned} \tag{2.15}$$

it holds $\frac{\mathrm{d}}{\mathrm{d}s}q(s) = 0$, which means that $q(s) = f(\gamma(s))$ is constant along the curve. Therefore, (2.14) can be solved by tracing back the characteristic curve and evaluating the corresponding initial data.

Often, the force term is neglected, i.e. $\mathbf{F} = 0$, which encounters the assumption that the acceleration is due only to external fields. Then the solution to the characteristic equations (2.15) reads

$$\begin{aligned}t(s) &= s + t_0,\\ \mathbf{x}(s) &= \mathbf{v}s + \mathbf{x}_0,\\ \mathbf{v}(s) &= \mathbf{v}_0\end{aligned} \tag{2.16}$$

for given initial $t_0 \geq 0$, $\mathbf{x}_0, \mathbf{v}_0 \in \mathbb{R}^3$. For ease, let $t_0 = 0$. The choice $s = t$ yields $f(\mathbf{x}(t), \mathbf{v}(t), t) = q(t)$ and especially $f(\mathbf{x}_0, \mathbf{v}_0, 0) = q(0)$. Since q is constant along solutions of (2.15), we obtain

$$f(\mathbf{x}, \mathbf{v}, t) = q(t) = q(0) = f(\mathbf{x}_0, \mathbf{v}_0, 0) = f^0(\mathbf{x}_0, \mathbf{v}_0) = f^0(\mathbf{x} - \mathbf{v}t, \mathbf{v}),$$

where we used (2.16). This solves the initial value problem (2.14) for $\mathbf{F} = 0$ uniquely according to the theorem of Picard-Lindelöf. And we conclude that the initial values $f_0(\mathbf{x})$ are transported along straight lines in direction of $\mathbf{v}$, see Figure 2.4.

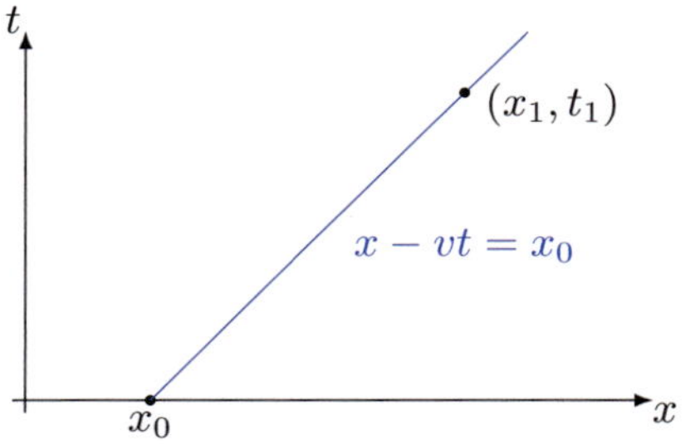

Figure 2.4: We illustrate the characteristic curve for the transport equation (2.14) with $\mathbf{F} = 0$. The solution at point (x_1, t_1) is given by the corresponding initial datum evaluated at $x_0 = x_1 - vt_1$ because solutions stay constant along characteristic curves.

In the framework of kinetic theory, the transport part (LHS of (2.5)) describes that particles move uniformly along straight lines in direction of $\mathbf{v}$. This observation matches Newton's first law 2.1.2. According to a possible nonvanishing force $\mathbf{F}$, the velocities of the particles can be deflected. Such forces often come by electric (and magnetic) fields for charged particles.

Coupling to electric field

Let us consider a charged particle within an electric field $\mathbf{E}$ and magnetic field $\mathbf{B}$. The so-called *Lorentz force* acts on this particle which is given by

$$\mathbf{F}_{\mathrm{L}} = Ze\,(\mathbf{E} + \mathbf{v} \times \mathbf{B})$$

with the charge state Z of the particle and the elementary charge e. The fields are externally given or provided by the solutions of Maxwell's equations [Max65]. These are the fundamental equations in electro-magneto dynamics, comparable to the role of Newton's laws in classical physics.

Axiom 2.1.24 (Maxwell's equations). *Maxwell's equations are the system of the following four PDEs.*

Gauß's law:	$\nabla_{\mathbf{x}} \cdot \mathbf{E} = \dfrac{q}{\varepsilon_0}$
Gauß's law for magnetism:	$\nabla_{\mathbf{x}} \cdot \mathbf{B} = 0$
Faraday's law of induction:	$\nabla_{\mathbf{x}} \times \mathbf{E} = -\partial_t \mathbf{B}$
Ampère's law with Maxwell's addition:	$\nabla_{\mathbf{x}} \times \mathbf{B} = \mu_0\,\mathbf{j} + \mu_0\varepsilon_0\,\partial_t \mathbf{E}$

Here, $\mathbf{E}(\mathbf{x},t)$ *denotes the electric field,* $\mathbf{B}(\mathbf{x},t)$ *the magnetic field,* $q(\mathbf{x},t)$ *the electric charge density,* $\mathbf{j}(\mathbf{x},t)$ *the electric current density,* ε_0 *the electric permittivity of vacuum and* μ_0 *the magnetic permeability of vacuum.*

In the context of kinetic theory, the electric charge and current density can be calculated via moments of the distribution functions.

Definition 2.1.25 (Electric charge and current density). *The electric charge and current density are given by*

$$q(\mathbf{x},t) = \sum_i Z_i e \int f_i(\mathbf{x},\mathbf{v},t)\,\mathrm{d}\mathbf{v} = \sum_i Z_i e n_i$$

$$\text{and} \quad \mathbf{j}(\mathbf{x},t) = \sum_i Z_i e \int \mathbf{v} f_i(\mathbf{x},\mathbf{v},t)\,\mathrm{d}\mathbf{v} = \sum_i Z_i e n_i \mathbf{u}_i$$

for $i \in S$.

We simplify Maxwell's equations in the following way [Son19]. We assume that the electric field represents the dominating term such that we neglect the magnetic field. Moreover, we suppose the fields to be quasi-static, and time derivatives will be neglected. We end up with the Lorentz force

$$\mathbf{F}_{\mathrm{L}} = Ze\,\mathbf{E} \tag{2.17}$$

where the electric field is determined by

$$\nabla_{\mathbf{x}} \times \mathbf{E} = 0 \tag{2.18}$$

$$\text{and} \quad \nabla_{\mathbf{x}} \cdot \mathbf{E} = \frac{q}{\varepsilon_0}. \tag{2.19}$$

We impose appropriate boundary conditions, e.g. periodic boundaries or the perfectly conducting boundary condition $\mathbf{E} \times \mathbf{n} = 0$ with the outer normal vector $\mathbf{n}$.

As the electric field is curl-free (2.18), there exists a scalar potential ϕ such that

$$\mathbf{E} = -\nabla_{\mathbf{x}}\phi. \tag{2.20}$$

Together with (2.19), this results in the *Poisson equation* for the potential:

$$-\Delta_{\mathbf{x}}\phi = \frac{q}{\varepsilon_0}. \tag{2.21}$$

We further assume that no external fields apply and that the electric field is self-consistent; that is the electric field represents a mean field acting on particles due to inhomogenities in the distribution functions.

In summary, the force $\mathbf{F}$ in (2.14) being responsible for the transport in velocity space is given by the simplified form of the Lorentz force (2.17). It originates from the self-consistent electric field which is determined by (2.20) and the Poisson equation (2.21).

The above set of equations describes interesting physical phenomena. In the following, we mention the Landau damping.

Landau damping We consider the interaction of particles via the self-consistent electric field. Any further type of collisions of the particles is excluded such that there will be no relaxation towards a thermal equilibrium. (This will be explained in the following sections.) We obtain a situation being described by the Vlasov-Poisson system. The distribution function is determined by

$$\partial_t f + \mathbf{v} \cdot \nabla_{\mathbf{x}} f + \frac{Ze}{m}\mathbf{E} \cdot \nabla_{\mathbf{v}} f = 0 \tag{2.22}$$

which is coupled to

$$\mathbf{E} = -\nabla_{\mathbf{x}}\phi \quad \text{and} \quad -\Delta_{\mathbf{x}}\phi = \frac{q}{\varepsilon_0}. \tag{2.23}$$

In one space dimension, (2.23) reduces to

$$\partial_x E = \frac{q}{\varepsilon_0}. \tag{2.24}$$

The Vlasov-Poisson system encounters a damping phenomenon of the electric field due to an initial disturbance from the equilibrium distribution. The so-called Landau damping is an irreversible process representing an uniformization in configuration space [Bal60]. It was mathematically predicted by Landau in 1946 [Lan46] for the linearized Vlasov-Poisson system and confirmed later by experiments. The mathematical result was made rigorous by Mouhot and Villani in 2010 [MV10, MV11].

For this phenomenon, we are interested in the distribution function of electrons f_e following (2.22). Ions are assumed to be immobile, and they are only considered as a neutralizing background density n_0 obeying the quasi-neutrality condition

$$0 = \int (n_0 - n_e)\,\mathrm{d}x = \int \left(n_0 - \int f_e \,\mathrm{d}\mathbf{v} \right) \mathrm{d}x.$$

The resulting PDE for the electric field is

$$\partial_x E = \frac{1}{\varepsilon_0}(n_0 - n_e) = \frac{1}{\varepsilon_0} \left(n_0 - \int f_e \,\mathrm{d}\mathbf{v} \right).$$

If collisions between particles are allowed and given e.g. by a BGK operator, the electric field is damped by two distinct effects [Bau21]: the Landau damping and the relaxation process according to the $\mathcal{H}$-Theorem which will be explained in the following section(s).

2.1.3 Boltzmann equation(s)

In 1872, Ludwig Boltzmann published a pioneering equation [Bol09]. It realistically describes phenomena in dilute atmosphere, e.g. aeronautics in high altitude or interactions in dilute plasmas, proven by experiments [Cer00, Vil02]. However, the equation was doubted by many physicists at that time because of the supposed contradiction between reversible microdynamics and the irreversibility related to the so-called $\mathcal{H}$-Theorem. Only after 100 years and many discussions, this concern could be resolved rigorously [Lan75]. Nowadays, the Boltzmann equation is fundamental in kinetic theory and celebrated in many articles. A lot of (introductory) articles and textbooks can be found in the literature [CC70, Cer88, Bab98, Cer00, Vil02, EP05, Gol06].

We make the following assumptions on a monatomic gas [Cer88, Vil02, GMM09]. Firstly, we assume a dilute gas such that only binary interactions take place. For N hard spheres with radius r in a three dimensional space, this translates to

$$Nr^3 << 1, \quad Nr^2 \approx 1.$$

Secondly, the description is based on classical mechanics without (chemical) reactions. Neither quantum nor relativistic effects are taken into account. This is sensible for a mean distance between particles which is larger than their thermal deBroglie wavelength and if the ratio of thermal speeds to the speed of light is small. References for extensions regarding the other cases are given in [Vil02, p. 17].

Thirdly, the collisions are localized in space and time. The duration time of an interaction is small compared to the typical time scales of the description.

Fourthly, we consider elastic collisions, so the microscopic conservation of momentum (2.2) and energy (2.3) are fulfilled.

Fifthly, the collisions are micro-reversible. This means that they are time reversible at the microscopic level. Speaking of statistics, the probability that $\mathbf{v}$ and $\mathbf{v}_*$ change into $\mathbf{v}'$ and $\mathbf{v}'_*$ due to a collision is as high as the probability that $\mathbf{v}'$ and $\mathbf{v}'_*$ change into $\mathbf{v}$ and $\mathbf{v}_*$.

Sixthly, we make Boltzmann's chaos assumption: Particles which are going to collide are uncorrelated. However, this assumption causes an asymmetry because particles are correlated after a collision. Actually, this asymmetry is crucial in order to explain the discrepancy between reversible microdynamics and irreversible mesodynamics which we discuss later.

With these assumptions, the Boltzmann equation can be formulated.

Definition 2.1.26 (Boltzmann equation). *The Boltzmann equation for a distribution function f reads*

$$\partial_t f + \mathbf{v} \cdot \nabla_{\mathbf{x}} f = \mathcal{Q}[f] \tag{2.25}$$

with the collision operator

$$\mathcal{Q}[f](\mathbf{x}, \mathbf{v}, t) \tag{2.26}$$

$$= \int_{\mathbb{R}^3} \int_{\mathcal{S}^2} (f(\mathbf{x}, \mathbf{v}', t) f(\mathbf{x}, \mathbf{v}'_*, t) - f(\mathbf{x}, \mathbf{v}, t) f(\mathbf{x}, \mathbf{v}_*, t)) K(|\mathbf{v} - \mathbf{v}_*|, \xi) \, \mathrm{d}\xi \, \mathrm{d}\mathbf{v}_*, \tag{2.27}$$

where $K(|\mathbf{v} - \mathbf{v}_*|, \xi)$ is the collision kernel, explained later, and ξ represents the unit vector in scattering direction.

This equation is often called Boltzmann transport equation, too.

In the following, we shortly motivate the collision operator. Let us have a look at two particles with velocities $\mathbf{v}$ and $\mathbf{v}_*$, respectively. The probability that these collide is given by

$$f(\mathbf{x}, \mathbf{v}, t) f(\mathbf{x}, \mathbf{v}_*, t),$$

where we use that particles are not correlated before and during the interactions. Because of the collision, the particles exchange momentum and energy such that the velocities convert into $\mathbf{v}'$ and $\mathbf{v}'_*$, respectively, see Figure 2.2. The relations of the velocities are predicted by the conservation of momentum and energy at the microscopic level (Theorems 2.1.6 and 2.1.7):

$$\begin{aligned} \mathbf{v} + \mathbf{v}_* &= \mathbf{v}' + \mathbf{v}'_* \\ \mathbf{v}^2 + \mathbf{v}_*^2 &= \mathbf{v}'^2 + \mathbf{v}'^2_*. \end{aligned}$$

This means that by such a collision the particles with velocities $\mathbf{v}$ and $\mathbf{v}_*$ get 'lost', whereas the particles with velocities $\mathbf{v}'$ and $\mathbf{v}'_*$ are 'gained'. The Boltzmann collision operator combines and sums up the effects of all possible interactions.

The details of these interactions are hidden in the collision kernel $K(|\mathbf{v}-\mathbf{v}_*|,\xi) = |\mathbf{v}-\mathbf{v}_*|\sigma$ with the differential cross section $\sigma = \sigma(|\mathbf{v}-\mathbf{v}_*|,\theta)$ and the deflection angle θ and $\cos(\theta) = \frac{\mathbf{v}-\mathbf{v}_*}{|\mathbf{v}-\mathbf{v}_*|}\cdot\xi$. A cross section can be seen as a probability that a collision occurs. Depending on the underlying model of the particles, several cross sections can be derived. The most common cross sections correspond to variable soft spheres, Maxwell molecules, pseudo-Maxwell molecules, hard spheres or Coulomb collisions (in plasmas). We give these examples in Section 2.2.3. The cross sections also have impacts on the needed regularity of f and its behavior at the tails [Vil02, p. 50].

The Boltzmann equation satisfies many physical properties such as conservation of mass, momentum and energy, a reasonable entropy behavior, and an $\mathcal{H}$-theorem can be proven.

We will specify these statements for the generalization to multi-species. Such an extension for gas mixtures can be found in [CC70, Cer88, Cer00].

Definition 2.1.27 (Multi-species Boltzmann equations). *The multi-species Boltzmann equations are a system of equations:*

$$\partial_t f_i + \mathbf{v}\cdot\nabla_{\mathbf{x}} f_i = \mathcal{Q}_i[\{f_i\}] \tag{2.28}$$

for $i \in S$ with the collision operator

$$\mathcal{Q}_i[\{f_i\}](\mathbf{x},\mathbf{v},t) = \sum_{j\in S}\mathcal{Q}_{ij}[f_i,f_j] \tag{2.29}$$

where the interactions between particles of species i and species j are described by

$$\mathcal{Q}_{ij}[f_i,f_j] \tag{2.30}$$

$$= \int_{\mathbb{R}^3}\int_{S^2}(f_i(\mathbf{x},\mathbf{v}',t)f_j(\mathbf{x},\mathbf{v}'_*,t) - f_i(\mathbf{x},\mathbf{v},t)f_j(\mathbf{x},\mathbf{v}_*,t))K_{ij}(|\mathbf{v}-\mathbf{v}_*|,\xi)\,\mathrm{d}\xi\,\mathrm{d}\mathbf{v}_* \tag{2.31}$$

with $K_{ij}(|\mathbf{v}-\mathbf{v}_*|,\xi) = |\mathbf{v}-\mathbf{v}_*|\sigma_{ij}$ and the differential cross section σ_{ij} for the species pair (i,j).

The pre- and post-collisional velocities follow again the microscopic conservation of momentum and energy (Theorems 2.1.6 and 2.1.7). But in the multi-species case, the masses need to be taken into account which results in

$$\begin{aligned} m_i\mathbf{v} + m_j\mathbf{v}_* &= m_i\mathbf{v}' + m_j\mathbf{v}'_* \\ m_i\mathbf{v}^2 + m_j\mathbf{v}_*^2 &= m_i\mathbf{v}'^2 + m_j\mathbf{v}'^2_*. \end{aligned}$$

The collision operator satisfies the following invariance properties which reflect important physical laws.

Theorem 2.1.28 (Conservation properties for multi-species Boltzmann equations). *The multi-species Boltzmann equations conserve mass, total momentum and total energy. This means for intra-species interactions, it holds*

$$\int \mathcal{Q}_{ii}[f_i, f_i] \, \mathrm{d}\mathbf{v} = 0, \tag{2.32}$$

$$\int \mathcal{Q}_{ii}[f_i, f_i] m_i \mathbf{v} \, \mathrm{d}\mathbf{v} = 0, \tag{2.33}$$

$$\int \mathcal{Q}_{ii}[f_i, f_i] m_i |\mathbf{v}|^2 \, \mathrm{d}\mathbf{v} = 0, \tag{2.34}$$

and for inter-species collisions, it holds

$$\int \mathcal{Q}_{ij}[f_i, f_j] \, \mathrm{d}\mathbf{v} = 0, \tag{2.35}$$

$$\int \mathcal{Q}_{ij}[f_i, f_j] m_i \mathbf{v} \, \mathrm{d}\mathbf{v} + \int \mathcal{Q}_{ji}[f_j, f_i] m_j \mathbf{v} \, \mathrm{d}\mathbf{v} = 0, \tag{2.36}$$

$$\int \mathcal{Q}_{ij}[f_i, f_j] m_i |\mathbf{v}|^2 \, \mathrm{d}\mathbf{v} + \int \mathcal{Q}_{ji}[f_j, f_i] m_j |\mathbf{v}|^2 \, \mathrm{d}\mathbf{v} = 0, \tag{2.37}$$

for $i, j \in S$ and $i \neq j$.

Proof. The proof can be found e.g. in [Gol06, Vil02, Pir18]. □

As already mentioned in the above theorem, these properties represent the conservation of mass, total momentum and total energy at the kinetic level. Since the kinetic regime is hard to imagine (f is not physically measurable), we illustrate this nomenclature in the following theorem.

Theorem 2.1.29 (Macroscopic equations for multi-species Boltzmann equations). *Let $f_1, f_2 \in L^\infty(\mathrm{d}\mathbf{v})$ decay fast enough to zero in $\mathbf{v}$. If f_1 and f_2 are solutions to* (2.28) *in the sense of distributions, the following local macroscopic equations are satisfied:*

$$\begin{aligned}
\partial_t n_i + \nabla_{\mathbf{x}} \cdot (n_i \mathbf{u}_i) &= 0, \\
\partial_t(\rho_i \mathbf{u}_i) + \nabla_{\mathbf{x}} \cdot \mathbb{P}_i + \nabla_{\mathbf{x}} \cdot (\rho_i \mathbf{u}_i \otimes \mathbf{u}_i) &= \sum_{\substack{j \in S \\ j \neq i}} \int m_i \mathcal{Q}_{ij}[f_i, f_j] \mathbf{v} \, \mathrm{d}\mathbf{v}, \\
\partial_t \left(\frac{\rho_i}{2} |\mathbf{u}_i|^2 + \frac{3}{2} n_i T_i \right) + \nabla_{\mathbf{x}} \cdot \mathbf{Q}_i &= \sum_{\substack{j \in S \\ j \neq i}} \int \frac{m_i}{2} \mathcal{Q}_{ij}[f_i, f_j] |\mathbf{v}|^2 \, \mathrm{d}\mathbf{v}.
\end{aligned}$$

Proof. We give the proof exemplary for the conservation of mass. The other equations are proven e.g. in [Pir18]. Under some integrability assumptions it holds

$$\begin{aligned}
0 &\overset{(2.32),(2.35)}{=} \int \mathcal{Q}_i[\{f_i\}] \, \mathrm{d}\mathbf{v} \overset{(2.28)}{=} \int [\partial_t f_i + \mathbf{v} \cdot \nabla_{\mathbf{x}} f_i] \, \mathrm{d}\mathbf{v} \\
&= \partial_t \int f_i \, \mathrm{d}\mathbf{v} + \nabla_{\mathbf{x}} \cdot \left(\int \mathbf{v} f_i \, \mathrm{d}\mathbf{v} \right) = \partial_t n_i + \nabla_{\mathbf{x}} \cdot (n_i \mathbf{u}_i).
\end{aligned}$$

□

Accordingly, the number density, respective mass density are transported through the space. A further integration with respect to $\mathbf{x}$ and Gauß's law yield that the total mass $\int \rho_i \,\mathrm{d}\mathbf{x}$ of species $i \in S$ can only change due to a flux over the boundary of the spatial domain. For proper boundary conditions, the flux vanishes and the total mass is conserved.

Similar statements can be formulated for the conservation of total momentum and the conservation of total energy. But in these cases, the species can exchange momentum and energy, which can be seen by the exchange terms on the RHS. Nevertheless, the sum over all species vanishes (due to Theorem 2.1.28), and total momentum and total energy are conserved for proper boundary conditions.

Moreover, the (multi-species) Boltzmann equation fulfils an $\mathcal{H}$-Theorem. This is a very fundamental property. In the beginning, it caused many doubts because of the resulting irreversibility at the mescoscopic level. But when the sceptics has been overcome, the relevance is tremendous. Before we discuss this in more detail, we define the entropy and corresponding quantities which the $\mathcal{H}$-Theorem is about.

Definition 2.1.30 (Entropy). *The entropy density functional of a gas is given by*

$$\begin{aligned}\mathcal{H}[\{f_i\}] &= \sum_{i\in S} \mathcal{H}_i[f_i] \\ \textit{with} \quad \mathcal{H}_i[f_i] &= \int f_i \log f_i \,\mathrm{d}\mathbf{v}\end{aligned} \tag{2.38}$$

and the entropy is

$$\int \mathcal{H}[\{f_i\}] \,\mathrm{d}\mathbf{x}. \tag{2.39}$$

The entropy dissipation functional reads

$$\begin{aligned}D &= \sum_{i\in S} D_i[\{f_i\}] \\ \textit{with} \quad D_i[\{f_i\}] &= \int \mathcal{Q}_i[\{f_i\}] \log f_i \,\mathrm{d}\mathbf{v},\end{aligned} \tag{2.40}$$

and we denote the relative entropy of f and g by

$$\mathcal{H}[f|g] = \int f \log \frac{f}{g} \,\mathrm{d}\mathbf{v}. \tag{2.41}$$

'Entropy' is also known in statistical mechanics, see (2.13); the Shannon entropy is a quantity of information [CT05]. The entropy (2.39) coincides with the Shannon entropy up to a sign. More comments on the physical content of (2.38) can be found in [Gra61, Vil02].

Often, the entropy *density* (2.38) is referred to entropy. Even more confusion comes up when in kinetic theory (and for hyperbolic systems of conservation laws) the notion of entropy is generally used to be a functional of specific use (see the $\mathcal{H}$-Theorem 2.1.31 below). So one better uses and thinks of 'entropy' deliberately.

We give the $\mathcal{H}$-Theorem here for the multi-species case. But the real work has been done by Boltzmann for the one-species model. The $\mathcal{H}$-Theorem was one of the most important contributions to statistical physics by Boltzmann [Vil02].

Theorem 2.1.31 ($\mathcal{H}$-Theorem for the Boltzmann equation). *The entropy is dissipated in the space homogeneous gas*

$$D \leq 0 \quad \Leftrightarrow \quad \partial_t \mathcal{H} \leq 0$$

with equality if and only if

$$f_i = \mathcal{M}[n_i, \mathbf{u}_{\mathrm{eq}}, T_{\mathrm{eq}}, m_i]$$

for $i \in S$ and an equilibrium velocity $\mathbf{u}_{\mathrm{eq}}$ and equilibrium temperature T_{eq} which is common to all species and specified in Proposition 2.1.33. The Maxwellian $\mathcal{M}$ is given in Definition 2.1.32.

In the space inhomogeneous case, it holds true

$$\partial_t \mathcal{H} + \nabla_{\mathbf{x}} \cdot \sum_{i \in S} \int \mathbf{v} f_i \log f_i \, \mathrm{d}\mathbf{v} \leq 0.$$

Proof. The proof can be found e.g. in [Cer00, paragraph 6.4]. It uses the basic inequality

$$(z - y) \log \left(\frac{y}{z} \right) \leq 0$$

for $y, z \in \mathbb{R}^+$ with equality if and only if $y = z$. Additionally, the proof relies on the so-called collision invariants and on the fact that from

$$\varphi(\mathbf{v}) + \varphi(\mathbf{v}_*) = \varphi(\mathbf{v}') + \varphi(\mathbf{v}'_*)$$

it follows that φ is a linear combination of 1, $\mathbf{v}$ and $|\mathbf{v}|^2$. □

Definition 2.1.32 (Maxwellian). *A Maxwellian $\mathcal{M}(\mathbf{x}, \mathbf{v}, t) \geq 0$ is a distribution function of the specific form*

$$\mathcal{M}(\mathbf{x}, \mathbf{v}, t) = \mathcal{M}[f](\mathbf{x}, \mathbf{v}, t) = \mathcal{M}[n, \mathbf{u}, T, m](\mathbf{x}, \mathbf{v}, t) = \frac{n}{\sqrt{2\pi \frac{T}{m}}^{\,d}} \exp\left(- \frac{|\mathbf{v} - \mathbf{u}|^2}{2 \frac{T}{m}} \right) \tag{2.42}$$

which shares the same macroscopic quantities with a corresponding distribution function f; that is

$$\begin{pmatrix} n \\ n\mathbf{u} \\ 3nT \end{pmatrix} = \int f \begin{pmatrix} 1 \\ \mathbf{v} \\ m|\mathbf{v} - \mathbf{u}|^2 \end{pmatrix} \mathrm{d}\mathbf{v}.$$

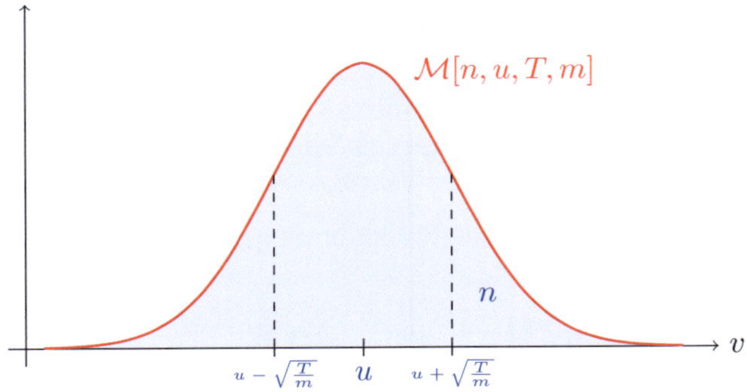

Figure 2.5: A Maxwellian $\mathcal{M}[n,u,T,m]$ is illustrated. This special Gaussian is centered around the mean velocity u, has integral n and standard deviation $\sqrt{\frac{T}{m}}$. Assuming a Maxwellian distribution, a particle has velocity $v \in \left[u-\sqrt{T/m}, u+\sqrt{T/m}\right]$ with around 68 %.

The exponent d corresponds to the dimensions in $\mathbf{v}$. In this thesis, it is $d=3$.
An illustration is given in Figure 2.5.

In the following, we discuss the meaning and consequences of the $\mathcal{H}$-Theorem and refer to [Vil02, Vil08].

The $\mathcal{H}$-Theorem states that the entropy $\mathcal{H}$ is nonincreasing. This is an exemplification of the second law of thermodynamics, given in Axiom 2.1.21. But, the second one being a postulate, the first one is even a theorem. Both describe the irreversibility of the macrodynamics. However, even though Clausius' law was accepted, the $\mathcal{H}$-Theorem gave rise to doubts since it was seen as a contradiction to the reversible mechanics the Boltzmann equation is derived from [CIP94, Leb95].

The irreversible macrodynamics do not contradict the reversible microdynamics provided that the macroscopic model is interpreted with the right amount of probability [EE07, Kac59] (a translated version of [EE07] can be found in [EEML60]). Referring to the Boltzmann equation, the probabilistic content sticks in the initial data: Among all microscopic configurations, which are compatible with the distribution function, we choose one to be the initial configuration. Hereby, the chaos assumption is relevant, and it is important to handle correctly that particles are uncorrelated before a collision, but they are correlated after the collision. Otherwise, contradictive statements can be constructed such as Loschmidt's paradox [Vil02]. Only in 1975, Lanford [Lan75] derived the Boltzmann equation rigorously from classical mechanics — at least for a perturbative framework and for hard-spheres collisions.

We now want to emphasize the importance of the $\mathcal{H}$-Theorem. From a physical point of view, we recognize the irreversibility of macrodynamics and the exemplification of the second law of thermodynamics. From a mathematical point of view, we make the following comments. To start with, we admit that the proof is not rigorous in a satisfactory generality because we do not know in general whether the solutions of Boltzmann's equation are smooth enough. However, the theorem's consequences and interpretations are pleasing.

There is a statistical (microscopic) meaning. The more exceptional an observed configuration is, the higher is the entropy. This fits the Shannon entropy which indicates how much information a signal contains.

Moreover, the $\mathcal{H}$-Theorem gives us powerful a priori estimates for the Boltzmann equation. It specifies that the entropy and also the entropy dissipation are finite, determined by the initial entropy.

The qualitative behavior for long times is also given by the $\mathcal{H}$-Theorem, and we can study the relaxation to equilibrium (when the collision term is zero). The entropy density decreases strictly and vanishes if and only if the distribution function has the specific form of a local Maxwellian. *Local* means that the corresponding number density, mean velocity and temperature may vary in $\mathbf{x}$. *Maxwellian* means yet that the distribution function behaves hydrodynamically. Together with the transport, the distribution function converges to a global Maxwellian; there is no spatial dependence anymore. To motivate this hydrodynamic limit mathematically, we bring some statements together: If the mean free path is short (the Knudsen number becomes small), the collisions play a very significant role. And then the finiteness of the entropy dissipation pushes the distribution function to be close to the local Maxwellian.

Both the $\mathcal{H}$-Theorem and the conservation properties are fundamental for the Boltzmann equation. They are also the elementary properties in order to call an approximative model *consistent*. Once, these properties are proven, several features follow automatically. In this sense, the conservation properties in Theorem 2.1.28 inherit that e.g. the mixture quantities in equilibrium can be precomputed from initial data in the space homogeneous case. The following statement is also proven for a more special case in [HHK+22].

Proposition 2.1.33. *In the spatially homogeneous case of a consistent kinetic equation, the mixture mean velocity* $\mathbf{u}_{\mathrm{mix}}$ *and mixture temperature* T_{mix}*, defined in Definition 2.1.16, stay constant in time.*

Proof. In the homogeneous setting, the number densities n_i, the mass densities ρ_i and the total momentum $\sum_{i\in S}\rho_i\mathbf{u}_i$ are all constant in time due to Theorem 2.1.28. Hence the ratio

$$\frac{\sum_{i\in S}\rho_i\mathbf{u}_i}{\sum_{i\in S}\rho_i}$$

which defines $\mathbf{u}_{\mathrm{mix}}$ is constant in time. To show that T_{mix} is constant in time, we use Definition 2.1.16 to write

$$\frac{3}{2}n_{\mathrm{mix}}T_{\mathrm{mix}} = \mathrm{I} - \mathrm{II}, \tag{2.43}$$

where

$$\mathrm{I} = \sum_{i\in S}\frac{1}{2}\rho_i|\mathbf{u}_i|^2 + \frac{3}{2}n_iT_i \tag{2.44}$$

is the total energy and

$$\mathrm{II} = \frac{1}{2}\rho_{\mathrm{mix}}|\mathbf{u}_{\mathrm{mix}}|^2. \tag{2.45}$$

In the homogeneous case, both I and II are constant in time. And we conclude that T_{mix} is also constant in time because of formula (2.43). □

Corollary 2.1.34. *In the homogeneous case of a consistent kinetic equation, the mean velocity $\mathbf{u}_{\mathrm{eq}}$ and temperature T_{eq} in equilibrium can be computed from the initial data by*

$$\mathbf{u}_{\mathrm{eq}} = \mathbf{u}_{\mathrm{mix}}(t=0) \quad \textit{and} \tag{2.46}$$

$$T_{\mathrm{eq}} = T_{\mathrm{mix}}(t=0) \tag{2.47}$$

where $\mathbf{u}_{\mathrm{mix}}$ and T_{mix} are the mixture quantities given in Definition 2.1.16.

Proof. This follows from Proposition 2.1.33 and the $\mathcal{H}$-Theorem 2.1.31 where it says that all species share the same mean velocity $\mathbf{u}_{\mathrm{eq}}$ and the same temperature T_{eq} in equilibrium. □

The formulae (2.46) and (2.47) apply to any kinetic model satisfying the conservation properties and the $\mathcal{H}$-Theorem. The latter one implies $|\mathbf{u}_i - \mathbf{u}_j|(t) \to 0$ for $t \to \infty$ such that, in equilibrium, the equation for the temperature simplifies to

$$T_{\mathrm{eq}} = T_{\mathrm{mix}}(t \to \infty) \overset{(2.9)}{=} \frac{\sum_i n_i T_i(t \to \infty)}{\sum_i n_i}.$$

For some models, even more can be said. E.g. for the multi-species BGK model in [KPP17, CKP20] it is known that the decay is exponentially fast, see Section 2.1.4.

The Boltzmann equation is fundamental in kinetic theory. But especially in numerics, this model suffers from some drawbacks. It is computationally costly to evaluate the complicated collision operator which makes implicit time discretizations extremely demanding. These are essential for strong interactions when approaching the hydrodynamic regime. So simplifications of the collision operator become necessary. One approach is the linearization, another approach is the modelling of the interaction part. The latter one may lead to the BGK equation.

2.1.4 BGK equation(s) and their extensions

In 1954, the so-called BGK model was published by Bathnagar, Gross and Krook [BGK54]. It is an approximation to the Boltzmann equation where (2.26) is replaced by a simpler operator. Basic properties like the conservation of mass, momentum and energy and an $\mathcal{H}$-theorem are verified. However, some properties are lost due to the approximation, e.g. the correct Prandtl number [Str97, ALPP00]. Nevertheless, in many occasions the benefits predominate, and the field of BGK equations is an active topic of research.

In the following, we motivate the BGK equation and refer to [BGK54, Str05]. The fundamental assumption for BGK models is that the distribution function is near equilibrium. Therefore, the distribution functions corresponding to the post-collisional velocities can be written as Maxwellians

$$
\begin{aligned}
&\int_{\mathbb{R}^3}\int_{\mathcal{S}^2} (f(\mathbf{v}')f(\mathbf{v}'_*) - f(\mathbf{v})f(\mathbf{v}_*))K(|\mathbf{v}-\mathbf{v}_*|,\xi)\,\mathrm{d}\xi\,\mathrm{d}\mathbf{v}_* \\
&\qquad \approx \int_{\mathbb{R}^3}\int_{\mathcal{S}^2} (\mathcal{M}(\mathbf{v}')\mathcal{M}(\mathbf{v}'_*) - f(\mathbf{v})f(\mathbf{v}_*))K(|\mathbf{v}-\mathbf{v}_*|,\xi)\,\mathrm{d}\xi\,\mathrm{d}\mathbf{v}_* \\
&\qquad = \int_{\mathbb{R}^3}\int_{\mathcal{S}^2} (\mathcal{M}(\mathbf{v})\mathcal{M}(\mathbf{v}_*) - f(\mathbf{v})f(\mathbf{v}_*))K(|\mathbf{v}-\mathbf{v}_*|,\xi)\,\mathrm{d}\xi\,\mathrm{d}\mathbf{v}_* \\
&\qquad = \mathcal{M}(\mathbf{v})\int_{\mathbb{R}^3}\int_{\mathcal{S}^2} \mathcal{M}(\mathbf{v}_*)K(|\mathbf{v}-\mathbf{v}_*|,\xi)\,\mathrm{d}\xi\,\mathrm{d}\mathbf{v}_* \\
&\qquad\quad - f(\mathbf{v})\int_{\mathbb{R}^3}\int_{\mathcal{S}^2} f(\mathbf{v}_*)K(|\mathbf{v}-\mathbf{v}_*|,\xi)\,\mathrm{d}\xi\,\mathrm{d}\mathbf{v}_*.
\end{aligned}
$$

The second but last equality holds because of

$$
\mathcal{M}(\mathbf{v}')\mathcal{M}(\mathbf{v}'_*) = \mathcal{M}(\mathbf{v})\mathcal{M}(\mathbf{v}_*)
$$

which comes by the microscopic conservation laws (2.2) and (2.3). Since we assume that we are close to equilibrium, we can state

$$
\int_{\mathbb{R}^3}\int_{\mathcal{S}^2} f(\mathbf{v}_*)K(|\mathbf{v}-\mathbf{v}_*|,\xi)\,\mathrm{d}\xi\,\mathrm{d}\mathbf{v}_* \approx \int_{\mathbb{R}^3}\int_{\mathcal{S}^2} \mathcal{M}(\mathbf{v}_*)K(|\mathbf{v}-\mathbf{v}_*|,\xi)\,\mathrm{d}\xi\,\mathrm{d}\mathbf{v}_*,
$$

and it follows

$$
\int_{\mathbb{R}^3}\int_{\mathcal{S}^2} (f(\mathbf{v}')f(\mathbf{v}'_*) - f(\mathbf{v})f(\mathbf{v}_*))K(|\mathbf{v}-\mathbf{v}_*|,\xi)\,\mathrm{d}\xi\,\mathrm{d}\mathbf{v}_* \approx \nu(\mathbf{v})(\mathcal{M}(\mathbf{v}) - f(\mathbf{v}))
$$

with the collision frequency

$$
\nu(\mathbf{x},\mathbf{v},t) = \int_{\mathbb{R}^3}\int_{\mathcal{S}^2} \mathcal{M}(\mathbf{x},\mathbf{v}_*,t)K(|\mathbf{v}-\mathbf{v}_*|,\xi)\,\mathrm{d}\xi\,\mathrm{d}\mathbf{v}_*. \tag{2.48}
$$

This integral (2.48) can be very complicated as discussed in Section 2.2.3. Only for very special cross sections (such as for Maxwellian molecules (2.90)), the integral is evaluable explicitely and independently of the microscopic velocity. Actually, a velocity-independent collision frequency simplifies the following calculations a lot. That is why usually the velocity dependence is avoided by very special cross sections, or an average value $\bar{\nu} = \frac{1}{n}\int \nu f\,\mathrm{d}\mathbf{v}$ is taken. However, this assumption includes physical drawbacks. In [Str97, HHK+21], the velocity challenge is faced.

In this section, we assume that the collision frequency is independent of the microscopic velocity; that is $\nu(\mathbf{x},\mathbf{v},t) = \nu(\mathbf{x},t)$. Then, we can introduce the following BGK equation.

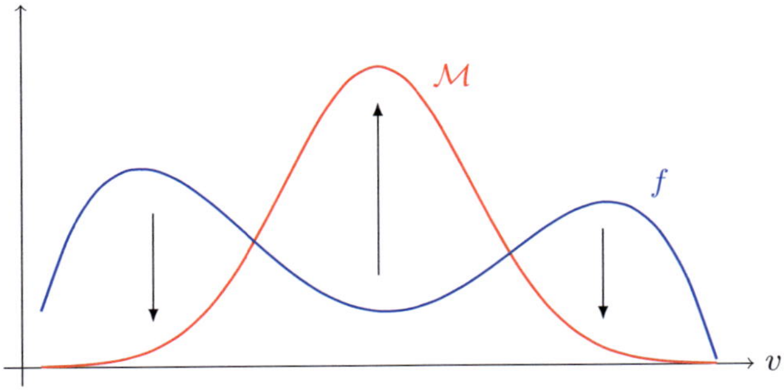

Figure 2.6: Illustration for the space homogeneous BGK equation: The distribution function f relaxes to the corresponding Maxwellian $\mathcal{M}$. When equilibrium is reached, it holds $f = \mathcal{M}$.

Definition 2.1.35 (BGK equation). *The BGK equation takes the form*

$$\partial_t f + \mathbf{v} \cdot \nabla_{\mathbf{x}} f = \mathcal{Q}[f] \tag{2.49}$$

with the BGK collision operator

$$\mathcal{Q}[f] = \nu(\mathcal{M}[f] - f) \tag{2.50}$$

and the collision frequency $\nu = \nu(\mathbf{x}, t)$.

In this context, the collision or interaction operator $\mathcal{Q}[f]$ is also called relaxation operator. This is motivated in the following and illustrated in Figure 2.6. In the space homogeneous case, the BGK equation reads

$$\partial_t f = \nu(\mathcal{M} - f).$$

If $\mathcal{M} > f$, then the RHS is positive. This results in a positive time derivative of f which means that f increases in time. However, if $\mathcal{M} < f$, then the negative RHS leads to a negative time derivative of f, and f decreases in time. In summary, f relaxes to $\mathcal{M}$.

Another way to motivate the BGK equation is a minimization procedure [ALPP00]. In a more general set-up, this is executed in Section 2.2, and the motivation for the standard BGK model is captured as special case for simplified collision frequencies.

The BGK equation is an approximation to the Boltzmann model. Yet, they share the same main properties.

Theorem 2.1.36 (Properties of the BGK model). *Let* $\nu, f > 0$ *and* $f \in L^\infty(\mathrm{d}\mathbf{v})$ *be a solution of the BGK equation decaying fast enough to zero for* $|\mathbf{v}| \to \infty$. *Then the BGK collision operator* (2.50) *shares the same main properties as the Boltzmann collision operator* (2.26)*, that is conservation of mass, momentum and energy,* $\mathcal{H}$*-Theorem and the structure of the equilibrium.*

Proof. The proof can be found in [Str05, Section 3.6]. The fact that the collision frequency is independent of the microscopic velocity is used in a fundamental way. We show this for the conservation of mass. It is

$$\int \mathcal{Q}[f]\,\mathrm{d}\mathbf{v} = \int \nu(\mathcal{M}[f]-f)\,\mathrm{d}\mathbf{v} \overset{(*)}{=} \nu \int (\mathcal{M}[f]-f)\,\mathrm{d}\mathbf{v} = 0.$$

The integral vanishes because the Maxwellian $\mathcal{M}[f]$ shares the same number density with f. However, we emphasize that $(*)$ only holds true for velocity-independent collision frequencies. □

Theorem 2.1.36 makes the BGK model to a consistent approximation of the Boltzmann equation. However, the simplification of the collision operator goes with some (physical) drawbacks. By extending the BGK model while maintaining a simplified RHS, one overcomes these physical discrepancies. In the following, we list some extensions which also can be combined. We do not claim completeness.

The Boltzmann equation reproduces the correct Prandtl number (Definition 2.1.19), whereas the standard BGK model gives $\mathrm{Pr} = 1$ which is unphysical. One way to fix this issue is to use ellipsoidal statistical BGK (ES-BGK) models [Hol66, ALPP00, KPP18, ZS05]. There, the Maxwellian in the relaxation term is replaced by a function of the same form but the scalar temperature is turned to a tensorial analogue. Alternatively, one may use a Shakhov model [Sha68, BY21] where the Maxwellian is multiplied with an extra term that adjusts the heat flux, but that does not affect the collision invariants.

For the one-species BGK model, the correct Prandtl number can also be regained by using velocity-dependent collision frequencies [Str97, MS04, ZS05]. For an extension to multi-species [HHK+21] one has not proven this additional feature, yet. Anyway, the velocity-dependent collision frequency is a way to come closer to the Boltzmann equation again because the constant collision frequency in the BGK model is an additional assumption.

We already presented the multi-species Boltzmann equations. This is a natural extension since, in nature, there are usually gas mixtures instead of pure one-species gases. For the same reasons, there exist several extensions of the BGK equation to multi species which can be devided into two different classes. The first class of multi-species BGK models uses one relaxation operator on the RHS [AAP02, BPS12, GMS11]. The second class of multi-species BGK models employs the same structure as the multi-species Boltzmann equations (2.28) which results in the sum of N relaxation operators for N species. We will present a general model of those in the following section. This model [KPP17] captures a lot of proposed models in the literature as special cases, such as [GK56, Ham65, Asi08, GSB89, SS01, Cer88, Gre73, HHM17a, BBG+18].

Another simplification used in the standard Boltzmann and BGK models is that only monatomic particles are considered. In order to deal with molecules, polyatomic models have been developed. The number of degrees of freedoms increases as rotations and vibrations become possible. We refer to the nice review in [Pir21].

In this thesis, we only consider physical phenomena. If chemical reactions should be taken into account, the model needs to be changed which especially influences the conservation of mass. References concerning with chemical reactive gases can be found in [BKPY21].

Currently, we only considered classical physics. But there are also extensions for relativistic and quantum regimes, respectively. BGK models including relativistic effects can be found in [PR18, BCNS12]. For quantum models, we refer to [BKPY21] and references therein. This is also concerned with in Section 2.3.

We are especially interested in gas mixtures, so we focus on models for multi species. In the following, we shortly recall the general consistent multi-species BGK model [KPP17]. It was developed by Klingenberg, Pirner and Puppo in 2017 which is why we use Klingenberg-Pirner-Puppo (KPP) in the following to refer to this model. Its special feature are free parameters which — appropriately chosen — generate many other multi-species models in the literature or which can be used to match additional physical properties.

KPP: a consistent multi-species BGK model

This model is nicely presented and many properties are summarized in [Pir18]. In the following, we give a short overview and also include more recent results.

For simplicity in notation and statements, we present the multi-species BGK model [KPP17] for two species, but the model can be extended to an arbitrary number of species because we only consider binary interactions. The structure of the RHS of the multi-species Boltzmann equations (2.28) is preserved, but every single collision operator is replaced by a BGK approximation consisting of a collision frequency $\nu_{kj} n_j$ multiplied by the deviation of the distributions from a local Maxwell distribution.

Definition 2.1.37 (Multi-species BGK equations). *The multi-species BGK equations in [KPP17] read*

$$\begin{aligned} \partial_t f_1 + \mathbf{v} \cdot \nabla_{\mathbf{x}} f_1 &= \mathcal{Q}_1[\{f_1, f_2\}], \\ \partial_t f_2 + \mathbf{v} \cdot \nabla_{\mathbf{x}} f_2 &= \mathcal{Q}_2[\{f_1, f_2\}] \end{aligned} \tag{2.51}$$

with the collision operators

$$\mathcal{Q}_1[\{f_1, f_2\}] = \mathcal{Q}_{11}[f_1, f_1] + \mathcal{Q}_{12}[f_1, f_2] = \nu_{11} n_1 (\mathcal{M}_{11} - f_1) + \nu_{12} n_2 (\mathcal{M}_{12} - f_1), \tag{2.52}$$

$$\mathcal{Q}_2[\{f_1, f_2\}] = \mathcal{Q}_{22}[f_2, f_2] + \mathcal{Q}_{21}[f_2, f_1] = \nu_{22} n_2 (\mathcal{M}_{22} - f_2) + \nu_{21} n_1 (\mathcal{M}_{21} - f_2) \tag{2.53}$$

and the collision frequencies per density ($i, j = 1, 2$)

$$\nu_{ij} = \nu_{ij}(\mathbf{x}, t).$$

The Maxwellians $\mathcal{M}_{11} = \mathcal{M}[f_1]$ and $\mathcal{M}_{22} = \mathcal{M}[f_2]$ are already known from the one-species case. On top of that, the mixture Maxwellians $\mathcal{M}_{12}$ and $\mathcal{M}_{21}$ are given by

$$\begin{aligned}
\mathcal{M}_{12}(\mathbf{x},\mathbf{v},t) &= \mathcal{M}[n_{12},\mathbf{u}_{12},T_{12},m_1] = \frac{n_{12}}{\sqrt{2\pi\frac{T_{12}}{m_1}}^3}\exp\left(-\frac{|\mathbf{v}-\mathbf{u}_{12}|^2}{2\frac{T_{12}}{m_1}}\right), \\
\mathcal{M}_{21}(\mathbf{x},\mathbf{v},t) &= \mathcal{M}[n_{21},\mathbf{u}_{21},T_{21},m_2] = \frac{n_{21}}{\sqrt{2\pi\frac{T_{21}}{m_2}}^3}\exp\left(-\frac{|\mathbf{v}-\mathbf{u}_{21}|^2}{2\frac{T_{21}}{m_2}}\right),
\end{aligned} \tag{2.54}$$

where n_{ij}, u_{ij} and T_{ij} still need to be defined. Actually, the collision frequencies $\nu_{ij}n_j$ and the macroscopic mixture quantities n_{ij}, u_{ij} and T_{ij} represent degrees of freedom which can be used in order to obtain desired properties. The main task is to satisfy the fundamental properties of the Boltzmann equation: conservation of mass, total momentum and total energy as well as the correct entropy behavior.

The collision frequencies $\nu_{11}n_1$ and $\nu_{22}n_2$ are related to intra-species collisions, whereas $\nu_{12}n_2$ and $\nu_{21}n_1$ correspond to inter-species collisions. Usually, the latter ones are linked by a constant, e.g. the mass ratio in the case of a plasma [Pir18, Section 5.2.1]. So we assume

$$\nu_{12} = \varepsilon\nu_{21}, \quad 0 < \varepsilon \leq 1. \tag{2.55}$$

The requirement on ε is not restrictive. If $\varepsilon > 1$, the exchange of the notation 1 and 2 yields $\frac{1}{\varepsilon} \leq 1$. In addition, all collision frequencies are assumed to be positive.

Remark 2.1.38. *In this section, the collision frequency is denoted by νn where ν is the collision frequency per density. This notation simplifies later statements and theorems. However, in the rest of this thesis the collision frequency itself is denoted by ν.*

The choices $\mathcal{M}_{11} = \mathcal{M}[f_1]$ and $\mathcal{M}_{22} = \mathcal{M}[f_2]$ guarantee the conservation of mass, momentum and energy during interactions of particles of one species with itself (2.32)–(2.34). Accordingly, the parameters $n_{12}, n_{21}, \mathbf{u}_{12}, \mathbf{u}_{21}, T_{12}$ and T_{21} for $\mathcal{M}_{12}$ and $\mathcal{M}_{21}$ are supposed to guarantee the conservation of the number of particles, total momentum and total energy (2.35)–(2.37) during inter-species interactions. We impose that $\mathbf{u}_{12}$ is a linear combination of $\mathbf{u}_1$ and $\mathbf{u}_2$, and T_{12} is a linear combination of T_1 and T_2 plus a velocity term. Then corresponding expression for $\mathbf{u}_{21}$ and T_{21} can be found while satisfying the conservation properties.

Theorem 2.1.39 (Conservation properties for the KPP model). *Let $\mathcal{M}_{11} = \mathcal{M}[f_1]$ and $\mathcal{M}_{22} = \mathcal{M}[f_2]$. For the mixture Maxwellians $\mathcal{M}_{12}(\mathbf{x},\mathbf{v},t) = \mathcal{M}[n_{12},\mathbf{u}_{12},T_{12},m_1]$ and $\mathcal{M}_{21}(\mathbf{x},\mathbf{v},t) = \mathcal{M}[n_{21},\mathbf{u}_{21},T_{21},m_2]$ define*

$$n_{12} = n_1 \quad \textit{and} \quad n_{21} = n_2, \tag{2.56}$$

and

$$\begin{aligned} \mathbf{u}_{12} &= \delta \mathbf{u}_1 + (1-\delta)\mathbf{u}_2, \\ \mathbf{u}_{21} &= \mathbf{u}_2 - \frac{m_1}{m_2}\varepsilon(1-\delta)(\mathbf{u}_2 - \mathbf{u}_1), \end{aligned} \tag{2.57}$$

as well as

$$\begin{aligned} T_{12} &= \alpha T_1 + (1-\alpha)T_2 + \gamma|\mathbf{u}_1 - \mathbf{u}_2|^2, \\ T_{21} &= \left[\frac{1}{3}\varepsilon m_1(1-\delta)\left(\frac{m_1}{m_2}\varepsilon(\delta-1)+\delta+1\right) - \varepsilon\gamma\right]|\mathbf{u}_1 - \mathbf{u}_2|^2 \\ &\quad + \varepsilon(1-\alpha)T_1 + (1-\varepsilon(1-\alpha))T_2, \end{aligned} \tag{2.58}$$

with

$$0 \leq \alpha \leq 1, \quad \delta \in \mathbb{R}, \quad \gamma \geq 0. \tag{2.59}$$

Then the BGK equations (2.51) *satisfy conservation of mass* (2.32)/(2.35), *total momentum* (2.33)/(2.36) *and total energy* (2.34)/(2.37).

Proof. The proofs can be found in [KPP17, Theorems 2.1–2.3]. □

The newly introduced parameters α, δ and γ play an essential role in this model. On the one hand, they can be physically interpreted. On the other hand, they equip the model with more degrees of freedom which can be used to match specific properties. Both features are discussed later. But first of all, we need to restrict δ and γ in order to ensure the positivity of temperatures.

Theorem 2.1.40 (Positivity of temperatures for the KPP model)**.** *All temperatures in* (2.51) *are positive if*

$$0 \leq \gamma \leq \frac{m_1}{3}(1-\delta)\left[(1+\frac{m_1}{m_2}\varepsilon)\delta + 1 - \frac{m_1}{m_2}\varepsilon\right], \tag{2.60}$$

and

$$\frac{\frac{m_1}{m_2}\varepsilon - 1}{1 + \frac{m_1}{m_2}\varepsilon} \leq \delta \leq 1. \tag{2.61}$$

Proof. The proof is given in [KPP17, Theorem 2.5]. □

Another fundamental property corresponds to the entropy.

Theorem 2.1.41 ($\mathcal{H}$-Theorem for the KPP model)**.** *Let* $f_1, f_2 > 0$. *Assume the relationships* (2.55)–(2.61) *for* $\alpha, \delta \neq 1$. *Then it holds*

$$\begin{aligned} D = &\int \nu_{11}(\mathcal{M}_{11} - f_1)\log f_1 \,\mathrm{d}\mathbf{v} + \int \nu_{12}(\mathcal{M}_{12} - f_1)\log f_1 \,\mathrm{d}\mathbf{v} \\ &+ \int \nu_{21}(\mathcal{M}_{21} - f_2)\log f_2 \,\mathrm{d}\mathbf{v} + \int \nu_{22}(\mathcal{M}_{22} - f_2)\log f_2 \,\mathrm{d}\mathbf{v} \leq 0 \end{aligned}$$

with equality if and only if f_1 *and* f_2 *are Maxwellians with equal mean velocity and temperature.*

Proof. The proof is given in [KPP17, Theorem 2.7]. □

As for the Boltzmann equation, the $\mathcal{H}$-Theorem gives us the qualitative behavior in equilibrium. There, the relaxation term vanishes which implies a specific form of the kernel — a Maxwellian.

Theoretical results of this model We give a short overview over recent theoretical results concerning existence of solutions and large-time behavior which is also given in [PWed].

In the periodic setting and under certain conditions on the initial data and the collision frequencies, there exist unique mild solutions to (2.51), proven in [KP17]. *Mild* solutions are a class of solutions with lower regularity than in the classical sense, comparable to the notion of *weak* solutions [Pir18].

In [BKYP22], there is another existence result concerning the existence of a unique global-in-time classical solution when the initial data are perturbed slightly from a global equilibrium.

Moreover, one can prove the following results on the large-time behavior.

Theorem 2.1.42 (Estimates for the distribution functions for the KPP model)**.** *Let ν_{12} be constant in time and consider the space homogeneous case. Then, we have the following decay rate of the distribution functions f_1 and f_2*

$$||f_i - \mathcal{M}_i||_{L^1(dv)} \leq 4e^{-\frac{1}{2}Ct}[\mathcal{H}(f_1^0|\mathcal{M}_1^0) + \mathcal{H}(f_2^0|\mathcal{M}_2^0)]^{\frac{1}{2}}, \quad i = 1, 2,$$

where C is the constant given by

$$C = \min\{\nu_{11}n_1 + \nu_{12}n_2, \nu_{22}n_2 + \nu_{21}n_1\}$$

and the superscript 0 denotes the value at time $t = 0$. Here, we use the relative entropy defined in (2.41).

Proof. The proof is given in [CKP20, Theorem 4.1]. □

Theorem 2.1.43 (Estimates for the mean velocities for the KPP model)**.** *Let ν_{12} be constant in time and consider the space homogeneous case. It holds the following relaxation rate*

$$\partial_t(\mathbf{u}_1 - \mathbf{u}_2) = \nu_{12}(1-\delta)\left(n_2 + \varepsilon\frac{m_1}{m_2}n_1\right)(\mathbf{u}_2 - \mathbf{u}_1) \tag{2.62}$$

and the following decay rate for the mean velocities

$$|\mathbf{u}_1(t) - \mathbf{u}_2(t)|^2 = e^{-2\nu_{12}(1-\delta)\left(n_2+\varepsilon\frac{m_1}{m_2}n_1\right)t}|\mathbf{u}_1(0) - \mathbf{u}_2(0)|^2. \tag{2.63}$$

Proof. The proof can be found in [CKP20, Theorem 4.2]. □

Theorem 2.1.44 (Estimates for the temperatures for the KPP model). *Let ν_{12} be constant in time and consider the space homogeneous case. It holds the following relaxation rate*

$$\partial_t(T_1 - T_2) = -C_1(T_1 - T_2) + C_2|\mathbf{u}_1 - \mathbf{u}_2|^2 \tag{2.64}$$

and the following decay rate for the temperatures

$$T_1(t) - T_2(t) = e^{-C_1 t}\left[T_1(0) - T_2(0) + \frac{C_2}{C_1 - C_3}(e^{(C_1-C_3)t} - 1)|\mathbf{u}_1(0) - \mathbf{u}_2(0)|^2\right], \tag{2.65}$$

where the constants are defined by

$$\begin{aligned}
C_1 &= (1-\alpha)\nu_{12}\,(n_2 + \varepsilon n_1)\,,\\
C_2 &= \nu_{12}\left(n_2\left((1-\delta)^2 + \frac{\gamma}{m_1}\right) - \varepsilon n_1\left(1 - \delta^2 - \frac{\gamma}{m_1}\right)\right),\\
C_3 &= 2\nu_{12}(1-\delta)\left(n_2 + \varepsilon\frac{m_1}{m_2}n_1\right).
\end{aligned}$$

Proof. The proof can be found in [CKP20, Theorem 4.3]. □

More results for the linearized collision operator in the space homogeneous case can be found in [LP19] where hypocoercivity in continuous phase space is studied. The authors prove exponential relaxation to global equilibrium with explicit rates. Their strategy is based on the entropy and spectral methods adapting Lyapunov's direct method as presented in [AAC16] for the one-species linearized BGK model.

In the hydrodynamic limit, we obtain the following result.

Theorem 2.1.45 (Macroscopic equations for the KPP model). *The macroscopic equations of the multi-species BGK model* (2.51) *read*

$$\begin{aligned}
\partial_t n_i + \nabla_x \cdot (n_i \mathbf{u}_i) &= 0,\\
\partial_t(\rho_i \mathbf{u}_i) + \nabla_x \cdot \mathbb{P}_i + \nabla_x \cdot (\rho_i \mathbf{u}_i \otimes \mathbf{u}_i) &= \mathbf{f}_{m_{i,j}},\\
\partial_t\left(\frac{\rho_i}{2}|\mathbf{u}_i|^2 + \frac{3}{2}n_i T_i\right) + \nabla_x \cdot Q_i &= F_{E_{i,j}},
\end{aligned}$$

with the exchange terms $\mathbf{f}_{m_{i,j}}$ and $F_{E_{i,j}}$ given by

$$\begin{aligned}
\mathbf{f}_{m_{1,2}} &= -\mathbf{f}_{m_{2,1}} = m_1\nu_{12}n_1n_2(1-\delta)(\mathbf{u}_2 - \mathbf{u}_1),\\
F_{E_{1,2}} &= -F_{E_{2,1}}\\
&= \left[\nu_{12}\frac{1}{2}n_1n_2m_1(\delta-1)(\mathbf{u}_1 + \mathbf{u}_2 + \delta(\mathbf{u}_1 - \mathbf{u}_2)) + \frac{1}{2}\nu_{12}n_1n_2\gamma(\mathbf{u}_1 - \mathbf{u}_2)\right]\\
&\qquad \cdot (\mathbf{u}_1 - \mathbf{u}_2) + \frac{3}{2}\varepsilon\nu_{21}n_1n_2(1-\alpha)(T_2 - T_1).
\end{aligned}$$

Proof. The LHS is the same as the one of the macroscopic equations for the Boltzmann equations in Theorem 2.1.29, but the RHS differs. For the Boltzmann operator, the integrals are only computable in special cases (e.g. for Maxwellian molecules, see (2.90)). Whereas for the BGK operators (2.52), it is possible to evaluate these integrals explicitly by using Theorem 2.1.39 and its proof. □

Meaning and possible choices of the free parameters In this section, we deal with the meaning and possible choices of the free parameters. In Theorems 2.1.43–2.1.45, we observe a physical meaning of α and δ. We see that α and δ show up in the exchange terms of momentum and energy as parameters in front of the relaxation of $\mathbf{u}_1$ towards $\mathbf{u}_2$ and T_1 towards T_2. So they determine, together with the collision frequencies, the speed of relaxation of the mean velocities and the temperatures to a common value.

These parameters can be tuned to make the model more physical. One possibility is to choose the parameters in such a way that the macroscopic exchange terms of momentum and energy match the ones for the Boltzmann equation. Since the Boltzmann equation is considered to be physical, the same applies in this case for the relaxation rates of the multi-species BGK equations.

The following calculations are performed by myself and published in [PWed]. We follow [HHM17a, Chapter 4.1] and compare the relaxation rates in the space homogeneous case to the relaxation rates for the space homogeneous Boltzmann equations. In [HHM17a], they find formulas for ν_{ij} such that either their relaxation rate for the mean velocities or their relaxation rate for the temperatures coincides with the corresponding rate of the Boltzmann equation. But using the free parameters α, δ and γ, we are able to match both of the relaxation rates at the same time.

The relaxation rates for the multi-species Boltzmann equations are given in [HHM17a] and read

$$\partial_t(\mathbf{u}_2 - \mathbf{u}_1)\Big|_{\text{Boltz}} = -\alpha_{12}\left(\frac{\rho_1+\rho_2}{\rho_1\rho_2}\cdot\frac{m_1+m_2}{2}\right)(\mathbf{u}_2-\mathbf{u}_1) \tag{2.66}$$

$$\partial_t(T_2 - T_1)\Big|_{\text{Boltz}} = -\alpha_{12}\left[\frac{n_1+n_2}{n_1 n_2}(T_2-T_1)+\frac{\rho_2-\rho_1}{3n_1n_2}|\mathbf{u}_2-\mathbf{u}_1|^2\right], \tag{2.67}$$

where α_{12} is a coefficient for energy transfer coming from the Boltzmann equation, see [HHM17a] and references therein. We compare the coefficients of the terms $\mathbf{u}_2 - \mathbf{u}_1$, $T_2 - T_1$ and $|\mathbf{u}_2 - \mathbf{u}_1|^2$ in these Boltzmann relaxation rates (2.66)–(2.67) with the coefficients for the BGK relaxation rates (2.62) and (2.64). This results in the specific parameters:

$$(\mathbf{u}_2 - \mathbf{u}_1)\text{-term:} \quad \delta = 1 - \frac{\alpha_{12}}{\nu_{12}} \frac{m_1 + m_2}{2} \frac{m_1 n_1 + m_2 n_2}{m_1 n_1 m_2 n_2} \left(n_1 \frac{m_1}{m_2} + n_2 \right)^{-1},$$

$$(T_2 - T_1)\text{-term:} \quad \alpha = 1 - \frac{\alpha_{12}}{\nu_{12} n_2 n_1},$$

$$|\mathbf{u}_2 - \mathbf{u}_1|^2\text{-term:}$$

$$\gamma = \frac{1}{3} (n_1 + n_2)^{-1} \left[\frac{\alpha_{12}}{\nu_{12}} \frac{m_2 n_2 - m_1 n_1}{n_2 n_1} - m_1 n_2 (1-\delta)^2 + m_1 n_1 (1 - \delta^2) \right].$$

Additionally, the constraints (2.59)–(2.61) need to be satisfied. This can be verified by a corresponding choice of ν_{ij}. One possibility is

$$\nu_{ij} = \frac{1}{2} \frac{\alpha_{ij}}{n_i n_j} \frac{(m_i + m_j)^2}{m_i m_j} \tag{2.68}$$

and for $1 \geq \varepsilon = \frac{m_i}{m_j}$ (cf. in a plasma). Then, ν_{11} and ν_{22} are still free and can be set in a desired way.

Alternatively but in an analogous way, the free parameters can be used to match the entropy decay rate or transport coefficients.

Another use for the parameters is the following one. They can be set such that we generate special cases of multi-species BGK models in the literature [GK56, Ham65, Asi08, GSB89, SS01, Cer88, Gre73, BBG+18, HHM17a].

All models, which can be considered as special cases of the class described here, enjoy the theoretical properties derived for the more general KPP model: an H-theorem, conservation properties, positivity of the inter-species temperatures and the quantitative decay rates to equilibrium.

If we exemplary choose

$$\varepsilon = 1, \quad \delta = \frac{m_1}{m_1 + m_2}, \quad \alpha = \frac{m_1^2 + m_2^2}{(m_1 + m_2)^2}, \quad \gamma = \frac{m_1 m_2}{(m_1 + m_2)^2} \frac{m_2}{3},$$

we obtain the model by Hamel in [Ham65]. This is done in more details in [KP17, Pir18].

In [HHM17a], Haack, Hauck and Murillo introduce their multi-species BGK model. They assume a symmetry of the mixed mean velocities $\mathbf{u}_{ij}$ and temperatures T_{ij} which coincides with the choice

$$\delta = \frac{m_1 \nu_{12}}{m_1 \nu_{12} + m_2 \nu_{21}}, \quad \alpha = \frac{\nu_{12}}{\nu_{12} + \nu_{21}}, \quad \gamma = \frac{m_1 m_2 \nu_{12} \nu_{21}}{3(\nu_{12} + \nu_{21})(m_1 \nu_{12} + \nu_{21} m_2)},$$

respective

$$\delta = \frac{m_1 \varepsilon}{m_2 + m_1 \varepsilon}, \quad \alpha = \frac{\varepsilon}{1 + \varepsilon}, \quad \gamma = \frac{\varepsilon m_1 m_2}{3(1+\varepsilon)(m_1 \varepsilon + m_2)}$$

provided that $\nu_{12} = \varepsilon\nu_{21}$. Be aware of the slightly different definition of ν_{ij} according to Remark 2.1.38. We used the notation of the KPP model here.

In their article [HHM17a], the authors additionally derive a fluid transport model by using a Chapman-Enskog expansion. On top of that, they illustrate the relaxation process in a spatially homogeneous plasma by numerical tests matching either the mean velocity or temperature relaxation rate to the Boltzmann one.

In [BBG+18], the authors derive a multi-species BGK model explicitely for an arbitrary number of species where they specify the free parameters for Maxwellian molecules. This notion refers to specific cross sections, see Section 2.2.3. It is

$$\delta = \frac{m_1}{m_1 + m_2}, \qquad \alpha = \frac{2\delta m_2}{m_1 + m_2}, \qquad \gamma = \frac{\delta m_2}{3}\left(\frac{2m_1}{m_1 + m_2} - \delta\right),$$

and the model [BBG+18] can be seen as a special case of the KPP model. This justifies in retrospect the linear combinations in KPP for the mixed quantities to be sensible and natural.

Having multi-species BGK models at hand, we aim to extend such a model by velocity-dependent collision frequencies.

2.2 Multi-species BGK models with velocity-dependent collision frequency

2.2.1 Motivation

Deriving the BGK equation (2.49) from the Boltzmann equation (2.25), the collision frequency ν appears as an integral over the differential cross section σ, see (2.48). The latter one defines the underlying physics during the collisions which is then inherited to the collision frequencies. In (2.48), we see that the collision frequency naturally depends on the microscopic velocity $\mathbf{v}$. However, in the standard BGK models and many extensions, this dependence is neglected for reasons of simplicity which we have comprehended in the proof of Theorem 2.1.36. Transport coefficients derived via the Chapman-Enskog expansion are very sensitive to the dynamics of the tails of the distribution function [CC70] so that this simplification may have profound effects on e.g. the resulting hydrodynamic behavior associated with this model. Now, equipped with around 50 years of further math experience, it makes sense to reintroduce the dependence on the microscopic velocity in order to come closer to the behavior of the Boltzmann equation again.

But also from physical considerations, we expect a velocity dependence. We can think of different motivations for respective types of interactions why the velocity dependence becomes important.

Firstly, we consider billiard balls. If all balls had exactly the same velocity, no collisions would occur at all. But the larger the relative velocity, the more collisions are expected, see Figure 2.7.

Figure 2.7: In both illustrations, the fluid's mean velocity is the same. In the left illustration, each particle's velocity coincides with the mean velocity so that the relative velocities vanish. Collisions between particles are not possible. However, in the right illustration, all particles have different velocities (in direction and absolute value) which results in very distinct relative velocities. This can lead to collisions.

We illustrate the second motivation in Figure 2.8. Especially for plasmas where a Coulomb force is encountered, it is better to think of interactions between particles than of collisions. Let us consider two particles in the frame of reference where particle 2 is at rest. The (effective) cross section in which the two particles interact is given by the dashed circle. Actually, the Coulomb potential ranges to infinity, but it is often replaced by an effective potential with finite range, see Section 2.2.3. If particle 1 moves fast (red line, for a large relative velocity), it spends little time in the cross section and just passes particle 2. The effect of an attractive interaction is small. Whereas, if particle 1 moves slowly (blue line, for a small relative velocity), it stays quite a while in the cross section such that the interaction has a large effect. We conclude that the microscopic velocity and more precise the relative velocity plays an important role for the interactions between particles.

Figure 2.8: An attractive interaction between particle 1 and particle 2 is illustrated. We assume the frame of reference where particle 2 is at rest and give the effective cross section (the area in which an interaction is possible) by a dashed circle. For a large velocity of particle 1 (red line), the effect of the interaction is small. But for a small velocity of particle 1 (blue line), particle 1 is deflected more due to a strong interaction.

Be aware that this is only an illustration. We still assume that collisions, respective interactions take place instantaneously. In our model, this means that these interactions à la Coulomb are condensed in time to a single moment. Alternatively speaking, the time scales for the interactions are much smaller than the system's time scale.

These two considerations show that there is a range of relative velocities where we really expect an impact on collisions. Since the temperature scales like $T \propto \int |\mathbf{v} - \mathbf{u}|^2 f \, \mathrm{d}\mathbf{v}$, which also involves the relative velocity, our observations translate to the temperature. One characteristic of a plasma is its usual high temperature. This explains why a velocity-dependent collision frequency might be more important in a plasma than in cold water.

2.2.2 A consistent BGK model with velocity-dependent collision frequency for gas mixtures

In the derivation of the standard BGK model, the collision frequency is assumed to be constant in velocity. This holds true only for very specific cross sections, but it simplifies the model a lot. The generalization to velocity-dependent collision frequencies poses many challenges in the structure and for the proofs.

Single-species BGK models with velocity-dependent collision frequency were introduced in [Cer66, BP93, Str97]. Since we are especially interested in gas mixtures, we recall the multi-species model presented in [HHK$^+$21] which includes [Str97] as a by-product.

For reasons of readability and simplicity in notation, we stick to two species. The extension to multi species can be found in [HHK$^+$21].

We consider two distribution functions $f_1 = f_1(\mathbf{x}, \mathbf{v}, t) \geq 0$ and $f_2 = f_2(\mathbf{x}, \mathbf{v}, t) \geq 0$ for the species with masses m_1 and m_2, respectively, and for $\mathbf{x} \in \mathbb{R}^3$ and $\mathbf{v} \in \mathbb{R}^3$ being the phase space variables and $t \geq 0$ the time.

Definition 2.2.1 (Multi-species BGK equations with velocity-dependent collision frequency). *The multi-species BGK equations in [HHK$^+$21] read*

$$
\begin{aligned}
\partial_t f_1 + \mathbf{v} \cdot \nabla_{\mathbf{x}} f_1 &= \mathcal{Q}_1[\{f_1, f_2\}], \\
\partial_t f_2 + \mathbf{v} \cdot \nabla_{\mathbf{x}} f_2 &= \mathcal{Q}_2[\{f_1, f_2\}]
\end{aligned}
\tag{2.69}
$$

with the collision operators

$$
\mathcal{Q}_1[\{f_1, f_2\}] = \mathcal{Q}_{11}[f_1, f_1] + \mathcal{Q}_{12}[f_1, f_2] = \nu_{11}(\mathcal{G}_{11} - f_1) + \nu_{12}(\mathcal{G}_{12} - f_1), \tag{2.70}
$$

$$
\mathcal{Q}_2[\{f_1, f_2\}] = \mathcal{Q}_{22}[f_2, f_2] + \mathcal{Q}_{21}[f_2, f_1] = \nu_{22}(\mathcal{G}_{22} - f_2) + \nu_{21}(\mathcal{G}_{21} - f_2) \tag{2.71}
$$

and the collision frequencies $(i, j = 1, 2)$

$$
\nu_{ij} = \nu_{ij}(\mathbf{x}, \mathbf{v}, t) \geq 0.
$$

In the following, we use the notation

$$
\mathbf{a}_i(\mathbf{v}) = m_i \begin{pmatrix} 1 \\ \mathbf{v} \\ |\mathbf{v}|^2 \end{pmatrix} \tag{2.72}
$$

for $i = 1, 2$. Before we specify the target functions $\mathcal{G}_{ij}$, we assume some integrability properties regarding the effect of the velocity-dependent collision frequencies.

Assumption 2.2.2. *Each collision frequency ν_{ij} is strictly positive with the additional property that*

$$\{\boldsymbol{\lambda} \in \mathbb{R}^5 \mid \exp(\boldsymbol{\lambda} \cdot \mathbf{a}_i) \in L^1(\mathbb{R}^3)\} = \{\boldsymbol{\lambda} = (\lambda^{(0)}, \boldsymbol{\lambda}^{(1)}, \lambda^{(2)}) \in \mathbb{R} \times \mathbb{R}^3 \times \mathbb{R} \mid \lambda^{(2)} < 0\}$$

is independent of i and j.

This assumption ensures integrability properties being satisfied for velocity-independent collision frequencies which holds true for many realistic frequency models.

The BGK system (2.69) maintains the structure from the multi-species Boltzmann (2.28) and BGK (2.51) equations. However, the velocity-dependent collision frequency introduces many challenges regarding the target functions $\mathcal{G}_{ij}$ which still need to be defined. They are chosen in such a way that they guarantee the consistency of the model by construction. This construction includes the constant collision frequency case in which the multi-species BGK model in [HHM17a] is regained.

We first give the form of the target functions $\mathcal{G}_{ij}$ followed by the interpretation and derivation.

Definition 2.2.3 (Target functions for the multi-species BGK model with $\nu(\mathbf{v})$). *The target functions for the multi-species BGK equations* (2.69) *are given by*

$$\mathcal{G}_{ii}[f_i] = e^{\boldsymbol{\lambda}_{ii} \cdot \mathbf{a}_i(\mathbf{v})}, \tag{2.73}$$

$$\mathcal{G}_{ij}[f_i, f_j] = e^{\boldsymbol{\lambda}_{ij} \cdot \mathbf{a}_i(\mathbf{v})} \tag{2.74}$$

for $i, j = 1, 2$, $i \neq j$. The parameters $\boldsymbol{\lambda}_{ii} = \left(\lambda_{ii}^{(0)}, \boldsymbol{\lambda}_{ii}^{(1)}, \lambda_{ii}^{(2)}\right)^\top \in \mathbb{R} \times \mathbb{R}^3 \times \mathbb{R}^-$ and $\boldsymbol{\lambda}_{ij} = \left(\lambda_{ij}^{(0)}, \boldsymbol{\lambda}^{(1)}, \lambda^{(2)}\right)^\top \in \mathbb{R} \times \mathbb{R}^3 \times \mathbb{R}^-$ depend on the corresponding distribution functions f_1 and f_2.

We can think of several names for $\mathcal{G}_{ij}$: target functions, attractors or generalized Maxwellians. This comes by different possibilities how these are interpreted. On the one hand, target function, respective attractor comes by the fact that the relaxation structure of (2.69) pushes f_i to $\mathcal{G}_{ii}$. On the other hand, since $\mathcal{G}_{ii}$ are Gaussian functions, for a given mass m there exist a number density $\tilde{n}$, a mean velocity $\tilde{\mathbf{u}}$, a temperature $\tilde{T}$ such that we can write

$$\mathcal{G}_{ii} = \mathcal{M}[\tilde{n}, \tilde{\mathbf{u}}, \tilde{T}, m].$$

However, this is missleading because in general this Maxwellian cannot be associated with the distribution function f_i, i.e. $\mathcal{G}_{ii} \neq \mathcal{M}[f_i]$. This is why we prefer the other names. In addition, we want to emphasize that, in general, we cannot express the macroscopic quantities n_i, $\mathbf{u}_i$ and T_i as analytic functions of $\boldsymbol{\lambda}_i$ and $\boldsymbol{\lambda}_{ij}$. For constant collision frequencies, this is possible (see Remark 3.2.3), but for velocity-dependent collision frequencies we do not know any explicit relationship. In the following section, we will see how $\mathcal{G}_{ii}$ and f_i yet are connected.

The form of these target functions actually comes by physical requirements. Let us first consider the *intra-species* target functions $\mathcal{G}_{ii}$.

Intra-species target functions

We seek a distribution function which minimizes the entropy while conserving mass, momentum and energy for intra-species interactions. Here we refer to 'entropy' to be a special functional. We weight the usual objective by the collision frequencies which in the end will yield the form of $\mathcal{G}_{ii}$. We define the strictly convex function

$$h(z) = z \log z - z, \quad z > 0,$$

and the constrained minimization problem reads

$$\min_{g \in \chi_{ii}} \int \nu_{ii} h(g) \,\mathrm{d}\mathbf{v}, \quad i \in \{1, 2\}, \tag{2.75}$$

where

$$\chi_{ii} = \left\{ g \;\middle|\; g > 0,\, \nu_{ii}(1 + |\mathbf{v}|^2) g \in L^1(\mathbb{R}^3),\, \int \nu_{ii} \mathbf{a}_i(\mathbf{v})(g - f_i) \,\mathrm{d}\mathbf{v} = 0 \right\}.$$

Via the constraints in the set χ_{ii}, the conservation properties (2.32)–(2.34) are ensured. The problem (2.75) can be reformulated by using the Lagrange functional

$$L_{ii}(g, \boldsymbol{\alpha}) = \int \nu_{ii} h(g) \,\mathrm{d}\mathbf{v} - \boldsymbol{\alpha} \cdot \int \nu_{ii} \mathbf{a}_i(\mathbf{v})(g - f_i) \,\mathrm{d}\mathbf{v}. \tag{2.76}$$

Any critical point satisfies the first-order optimality condition

$$\frac{\delta L_{ii}}{\delta g}(\mathcal{G}_{ii}, \boldsymbol{\lambda}_{ii}) = \nu_{ii}(\ln \mathcal{G}_{ii} - \boldsymbol{\lambda}_{ii} \cdot \mathbf{a}_i(\mathbf{v})) = 0,$$

which results in

$$\mathcal{G}_{ii} = \exp\left(\boldsymbol{\lambda}_{ii} \cdot \mathbf{a}_i(\mathbf{v})\right) = \exp\left(m_i \lambda_{ii}^{(0)} + m_i \boldsymbol{\lambda}_{ii}^{(1)} \cdot \mathbf{v} + m_i \lambda_{ii}^{(2)} |\mathbf{v}|^2 \right).$$

For $\Lambda := \{\boldsymbol{\alpha} = (\alpha^{(0)}, \boldsymbol{\alpha}^{(1)}, \alpha^{(2)})^\top \in \mathbb{R} \times \mathbb{R}^3 \times \mathbb{R}^-\}$, the multipliers $\boldsymbol{\lambda}_{ii} \in \Lambda$ solve the dual of (2.76):

$$\boldsymbol{\lambda}_{ii} = \operatorname*{argmin}_{\boldsymbol{\alpha} \in \Lambda} z_{ii}(\boldsymbol{\alpha}, \boldsymbol{\mu}_{ii}) \tag{2.77}$$

with

$$z_{ii}(\boldsymbol{\alpha}, \boldsymbol{\mu}_{ii}) = -\int \nu_{ii} e^{\boldsymbol{\alpha} \cdot \mathbf{a}_i} \,\mathrm{d}\mathbf{v} + \boldsymbol{\alpha} \cdot \boldsymbol{\mu}_{ii}$$

and

$$\boldsymbol{\mu}_{ii} = \int \nu_{ii} \mathbf{a}_i f_i \, \mathrm{d}\mathbf{v}.$$

Having motivated the intra-species target functions, we prove that they are well-posed. For this, we define the set

$$I_{ij} := \{g \geq 0 \,|\, \nu_{ij}(1 + |\mathbf{v}|^2)g \in L^1(\mathbb{R}^3), g \not\equiv 0\}$$

clarifying integrability conditions.

Theorem 2.2.4 (Well-posedness of intra-species target functions for the multi-species BGK model with $\nu(\mathbf{v})$). *There exists a unique function of the form* (2.73) *which minimizes* (2.75) *for any* $f_i \in I_{ii}$.

Proof. The entire proof is given in [HHK+21, Theorems 1 and 3]. We shortly summarize the basic ideas. Firstly, suppose that there exists a $\boldsymbol{\lambda}_{ii} \in \Lambda$ such that $\mathcal{G}_{ii} = e^{\boldsymbol{\lambda}_{ii} \cdot \mathbf{a}} \in \chi_{ii}$. Then the convexity of h and $h'(\mathcal{G}_{ii}) = \boldsymbol{\lambda}_{ii} \cdot \mathbf{a}_i$ lead to

$$h(g) \geq h(\mathcal{G}_{ii}) + \boldsymbol{\lambda}_{ii} \cdot \mathbf{a}_i(g - \mathcal{G}_{ii}).$$

Because of $\nu_{ii} \geq 0$, it holds for all $g \in \chi_{ii}$

$$\int \nu_{ii} h(g) \, \mathrm{d}\mathbf{v} \geq \int \nu_{ii} h(\mathcal{G}_{ii}) \, \mathrm{d}\mathbf{v} + \boldsymbol{\lambda}_{ii} \cdot \int \nu_{ii} \mathbf{a}_i (g - \mathcal{G}_{ii}) \, \mathrm{d}\mathbf{v} = \int \nu_{ii} h(\mathcal{G}_{ii}) \, \mathrm{d}\mathbf{v}.$$

So $\mathcal{G}_{ii}$ is a minimizer of (2.75), and due to the strict convexity of h, $\mathcal{G}_{ii}$ is unique.

The second step is to show that there always exists such a unique $\boldsymbol{\lambda}_{ii} \in \Lambda$ for any $f_i \in I_{ii}$. This was rigorously proven in [HHK+21]. Basically, the authors show that the dual function $z_{ii}(\boldsymbol{\alpha}, \boldsymbol{\mu})$ is differentiable and that it attains its unique minimum at $\boldsymbol{\lambda}_{ii}$ on Λ with $\exp(\boldsymbol{\lambda}_{ii} \cdot \mathbf{a}_i) \in I_{ii}$ for any $\boldsymbol{\mu}$ of the form $\int \nu_{ii} \mathbf{a}_i g(\mathbf{v}) \, \mathrm{d}\mathbf{v}$ and $g \in I_{ii}$. A main ingredient, following [Jun00], is to prove that z_{ii} attains a minimum along the line $\boldsymbol{\alpha} + s\boldsymbol{\xi}$ in any direction $\boldsymbol{\xi} \in S^5$, and these minima are not located at the boundary. Using the implicit function theorem, this statement can be tied up for the entire function $z_{ii}(\boldsymbol{\alpha}, \boldsymbol{\mu})$ which results in one minimum of z_{ii}. Then, for the thus found minimum, the necessary condition yields

$$0 = \nabla_{\boldsymbol{\alpha}} z_{ii}(\boldsymbol{\lambda}_{ii}, \boldsymbol{\mu})$$

giving $\boldsymbol{\mu} = \int \nu_{ii} \mathbf{a}_i e^{\boldsymbol{\lambda}_{ii} \cdot \mathbf{a}_i} \, \mathrm{d}\mathbf{v}$. The result follows from the first part of the proof. □

We continue with inter-species interactions.

Inter-species target functions

For the inter-species interactions, we seek a solution to

$$\min_{g_1, g_2 \in \chi_{12}} \int \nu_{12} h(g_1) \, \mathrm{d}\mathbf{v} + \int \nu_{21} h(g_2) \, \mathrm{d}\mathbf{v}, \tag{2.78}$$

where

$$\chi_{12} = \Bigg\{ (g_1, g_2) \,\Big|\, g_1, g_2 > 0,\ \nu_{12}(1+|\mathbf{v}|^2) g_1,\ \nu_{21}(1+|\mathbf{v}|^2) g_2 \in L^1(\mathbb{R}^3),$$
$$\int m_1 \nu_{12} (g_1 - f_1)\, \mathrm{d}\mathbf{v} = 0, \quad \int m_2 \nu_{21} (g_2 - f_2)\, \mathrm{d}\mathbf{v} = 0,$$
$$\int m_1 \nu_{12} \begin{pmatrix} \mathbf{v} \\ |\mathbf{v}|^2 \end{pmatrix} (g_1 - f_1)\, \mathrm{d}\mathbf{v} + \int m_2 \nu_{21} \begin{pmatrix} \mathbf{v} \\ |\mathbf{v}|^2 \end{pmatrix} (g_2 - f_2)\, \mathrm{d}\mathbf{v} = 0 \Bigg\}. \tag{2.79}$$

The set χ_{12} ensures that the mass for each species, the total momentum and the total energy for inter-species collisions are conserved (2.35)–(2.37). The corresponding Lagrange functional reads

$$\begin{aligned} &L_{12} : \chi_{12} \times \mathbb{R} \times \mathbb{R} \times \mathbb{R}^3 \times \mathbb{R} \to \mathbb{R}, \\ &L_{12}(g_1, g_2, \alpha_{12}^{(0)}, \alpha_{21}^{(0)}, \boldsymbol{\alpha}^{(1)}, \alpha^{(2)}) = \int \nu_{12} (g_1 \ln g_1 - g_1)\, \mathrm{d}\mathbf{v} + \int \nu_{21} (g_2 \ln g_2 - g_2)\, \mathrm{d}\mathbf{v} \\ &\qquad - \alpha_{12}^{(0)} \int m_1 \nu_{12} (g_1 - f_1)\, \mathrm{d}\mathbf{v} - \alpha_{21}^{(0)} \int m_2 \nu_{21} (g_2 - f_2)\, \mathrm{d}\mathbf{v} \\ &\qquad - \boldsymbol{\alpha}^{(1)} \cdot \left(\int m_1 \nu_{12} \mathbf{v} (g_1 - f_1)\, \mathrm{d}\mathbf{v} + \int m_2 \nu_{21} \mathbf{v} (g_2 - f_2)\, \mathrm{d}\mathbf{v} \right) \\ &\qquad - \alpha^{(2)} \left(\int m_1 \nu_{12} |\mathbf{v}|^2 (g_1 - f_1)\, \mathrm{d}\mathbf{v} + \int m_2 \nu_{21} |\mathbf{v}|^2 (g_2 - f_2)\, \mathrm{d}\mathbf{v} \right). \end{aligned} \tag{2.80}$$

The first-order optimality conditions

$$\begin{aligned} \frac{\delta L_{12}}{\delta g_1}(\mathcal{G}_{12}, \mathcal{G}_{21}, \lambda_{12}^{(0)}, \lambda_{21}^{(0)}, \boldsymbol{\lambda}^{(1)}, \lambda^{(2)}) &= \nu_{12} (\ln \mathcal{G}_{12} - \boldsymbol{\lambda}_{12} \cdot \mathbf{a}_1) = 0, \\ \frac{\delta L_{12}}{\delta g_2}(\mathcal{G}_{12}, \mathcal{G}_{21}, \lambda_{12}^{(0)}, \lambda_{21}^{(0)}, \boldsymbol{\lambda}^{(1)}, \lambda^{(2)}) &= \nu_{21} (\ln \mathcal{G}_{21} - \boldsymbol{\lambda}_{21} \cdot \mathbf{a}_2) = 0 \end{aligned}$$

need to be satisfied by any critical point $(\mathcal{G}_{12}, \mathcal{G}_{21}, \lambda_{12}^{(0)}, \lambda_{21}^{(0)}, \boldsymbol{\lambda}^{(1)}, \lambda^{(2)})$. We end up with

$$\begin{aligned} \mathcal{G}_{12} &= \exp(\boldsymbol{\lambda}_{12} \cdot \mathbf{a}_1) = \exp\left(m_1 \lambda_{12}^{(0)} + m_1 \boldsymbol{\lambda}^{(1)} \cdot \mathbf{v} + m_1 \lambda^{(2)} |\mathbf{v}|^2 \right), \\ \mathcal{G}_{21} &= \exp(\boldsymbol{\lambda}_{21} \cdot \mathbf{a}_2) = \exp\left(m_2 \lambda_{21}^{(0)} + m_2 \boldsymbol{\lambda}^{(1)} \cdot \mathbf{v} + m_2 \lambda^{(2)} |\mathbf{v}|^2 \right). \end{aligned}$$

We only require conservation of the combined momentum and energy. This leads to only one Lagrange multiplier for the momentum constraint and one Lagrange multiplier for the energy constraint. Therefore, $\boldsymbol{\lambda}^{(1)}$ and $\lambda^{(2)}$ are the same for both $\mathcal{G}_{12}$ and $\mathcal{G}_{21}$. When the collision frequency is constant, this restriction is the same as the one used in [HHM17a], but more restrictive than in model [KPP17].

If $\Lambda_{12} = \{(\boldsymbol{\alpha}_{12}, \boldsymbol{\alpha}_{21}) : \boldsymbol{\alpha}_{ij} = (\alpha_{ij}^{(0)}, \boldsymbol{\alpha}^{(1)}, \alpha^{(2)})^\top \in \mathbb{R} \times \mathbb{R}^3 \times \mathbb{R}^-\}$, then the multipliers $(\boldsymbol{\lambda}_{12}, \boldsymbol{\lambda}_{21})$ solve the dual problem of (2.80):

$$
\begin{aligned}
(\boldsymbol{\lambda}_{12}, \boldsymbol{\lambda}_{21}) = \underset{(\boldsymbol{\alpha}_{12}, \boldsymbol{\alpha}_{21}) \in \Lambda_{12}}{\operatorname{argmin}} \bigg\{ & - \int (\nu_{12} e^{\boldsymbol{\alpha}_{12} \cdot \mathbf{a}_1} + \nu_{21} e^{\boldsymbol{\alpha}_{21} \cdot \mathbf{a}_2}) \, \mathrm{d}\mathbf{v} \\
& + \alpha_{12}^{(0)} \int m_1 \nu_{12} f_1 \, \mathrm{d}\mathbf{v} + \alpha_{21}^{(0)} \int m_2 \nu_{21} f_2 \, \mathrm{d}\mathbf{v} \\
& + \boldsymbol{\alpha}^{(1)} \cdot \int \mathbf{v} (m_1 \nu_{12} f_1 + m_2 \nu_{21} f_2) \, \mathrm{d}\mathbf{v} \\
& + \alpha^{(2)} \int |\mathbf{v}|^2 (m_1 \nu_{12} f_1 + m_2 \nu_{21} f_2) \, \mathrm{d}\mathbf{v} \bigg\}.
\end{aligned}
\tag{2.81}
$$

We now consider the well-posedness of $\mathcal{G}_{12}$ and $\mathcal{G}_{21}$.

Theorem 2.2.5 (Well-posedness of inter-species target functions for the multi-species BGK model with $\nu(\mathbf{v})$). *There exists a unique solution of the form* (2.74) *which minimizes* (2.78) *for any* $f_i \in I_{ij}$.

Proof. The proof is given in [HHK+21, Theorems 2 and 4]. The ideas of Theorem 2.2.4 are essentially extended to the mixture case. □

After having established the model, we can discuss its properties.

Consistency of the model

The consistency of the presented model is given in the following theorems.

Theorem 2.2.6 (Conservation properties for the multi-species BGK model with $\nu(\mathbf{v})$). *The model* (2.69) *conserves mass, total momentum and total energy; that is*

$$
\int \nu_{ii} (\mathcal{G}_{ii} - f_i) \mathbf{a}_i \, \mathrm{d}\mathbf{v} = 0 \quad \textit{for} \quad i = 1, 2 \quad \textit{and} \tag{2.82}
$$

$$
\int \nu_{12} (\mathcal{G}_{12} - f_1) \mathbf{a}_1 \, \mathrm{d}\mathbf{v} + \int \nu_{21} (\mathcal{G}_{21} - f_2) \mathbf{a}_2 \, \mathrm{d}\mathbf{v} = 0. \tag{2.83}
$$

Proof. This is an immediate consequence of the construction of the target functions. □

In a kinetic context, the conservation of mass, momentum and energy are associated with the equations (2.82) and (2.83). However, it is useful to keep in mind that these correspond to

$$\partial_t \rho_1 + \nabla_x \cdot \int m_1 \mathbf{v} f_1 \,\mathrm{d}\mathbf{v} = 0, \tag{2.84a}$$

$$\partial_t \rho_2 + \nabla_x \cdot \int m_2 \mathbf{v} f_2 \,\mathrm{d}\mathbf{v} = 0, \tag{2.84b}$$

$$\partial_t (\rho_1 \mathbf{u}_1 + \rho_2 \mathbf{u}_2) + \nabla_x \cdot \left(\int \mathbf{v} \otimes \mathbf{v} (m_1 f_1 + m_2 f_2) \,\mathrm{d}\mathbf{v} \right) = 0, \tag{2.84c}$$

$$\partial_t \left(\frac{\rho_1 |\mathbf{u}_1|^2}{2} + \frac{3\rho_1 T_1}{2m_1} + \frac{\rho_2 |\mathbf{u}_2|^2}{2} + \frac{3\rho_2 T_2}{2m_2} \right) + \nabla_x \cdot \left(\int \mathbf{v} \frac{|\mathbf{v}|^2}{2} (m_1 f_1 + m_2 f_2) \,\mathrm{d}\mathbf{v} \right) = 0. \tag{2.84d}$$

The equations (2.84) illustrate the conservation properties and are also common in the hydrodynamic context.

We continue with the entropy behavior of the model.

Theorem 2.2.7 ($\mathcal{H}$-Theorem for the multi-species BGK model with $\nu(\mathbf{v})$)**.** *Assume $f_1, f_2 > 0$ and the validity of* (2.82) *and* (2.83). *Let us denote the function $h(z) = z\ln(z) - z$ and the total entropy $\mathcal{H}[\{f_1, f_2\}] = \int (h(f_1) + h(f_2)) \,\mathrm{d}\mathbf{v}$. Then, the following entropy inequality holds true*

$$\partial_t \mathcal{H}[\{f_1, f_2\}] + \nabla_{\mathbf{x}} \cdot \left[\int \mathbf{v} (h(f_1) + h(f_2)) \,\mathrm{d}\mathbf{v} \right] \leq 0 \tag{2.85}$$

with equality if and only if f_1 and f_2 are two Maxwell distributions with equal mean velocity and temperature.

Proof. The proof can be found in [HHK+21, Theorem 6 and Corollary 3] which uses standard techniques regarding entropy. For convenience, we recall it here. A direct calculation gives

$$\begin{aligned} \partial_t \mathcal{H}[\{f_1, f_2\}] + \nabla_{\mathbf{x}} \cdot \left[\int \mathbf{v} (h(f_1) + h(f_2)) \,\mathrm{d}\mathbf{v} \right] &= \sum_{i=1}^{2} \int \mathcal{Q}_i[\{f_i\}] \log f_i \,\mathrm{d}\mathbf{v} \\ &= D_{11} + D_{12} + D_{21} + D_{22} \end{aligned}$$

with the dissipation terms

$$D_{ij} = \int \nu_{ij} (\mathcal{G}_{ij} - f_i) \log f_i \,\mathrm{d}\mathbf{v}.$$

We show that $D_{ii} \leq 0$ and $D_{12} + D_{21} \leq 0$ which proves (2.85).

Due to the conservation during intra-species interactions (2.82), it is

$$\int \nu_{ii} (\mathcal{G}_{ii} - f_i) \log \mathcal{G}_{ii} \,\mathrm{d}\mathbf{v} = \boldsymbol{\lambda}_{ii} \cdot \int \nu_{ii} (\mathcal{G}_{ii} - f_i) \,\mathbf{a}_i \,\mathrm{d}\mathbf{v} = 0.$$

It follows

$$D_{ii} = \int \nu_{ii}(\mathcal{G}_{ii} - f_i) \log f_i \, \mathrm{d}\mathbf{v} - \int \nu_{ii}(\mathcal{G}_{ii} - f_i) \log \mathcal{G}_{ii} \, \mathrm{d}\mathbf{v}$$
$$= \int \nu_{ii}(\mathcal{G}_{ii} - f_i) \log \frac{f_i}{\mathcal{G}_{ii}} \, \mathrm{d}\mathbf{v} \leq 0 \tag{2.86}$$

where the last inequality holds because

$$(z - y) \log \left(\frac{y}{z}\right) \leq 0 \tag{2.87}$$

for $y, z \in \mathbb{R}^+$. Additionally, (2.87) is an equality if and only if $y = z$. Applied to (2.86), $D_{ii} = 0$ if and only if $\mathcal{G}_{ii} = f_i$. In that case, f_i is a Gaussian function, and since $\mathcal{G}_{ii}$ and f_i obviously share the same moments, $f_i = \mathcal{M}[f_i]$ is even a Maxwellian.

For the mixed dissipation terms, we use the conservation during inter-species interactions (2.83) and observe

$$\begin{aligned}
I := & \int \nu_{12}(\mathcal{G}_{12} - f_1) \log \mathcal{G}_{12} \, \mathrm{d}\mathbf{v} + \int \nu_{21}(\mathcal{G}_{21} - f_2) \log \mathcal{G}_{21} \, \mathrm{d}\mathbf{v} \\
= & \ \lambda_{12}^{(0)} \int \nu_{12}(\mathcal{G}_{12} - f_1) m_1 \, \mathrm{d}\mathbf{v} + \lambda_{21}^{(0)} \int \nu_{21}(\mathcal{G}_{21} - f_2) m_2 \, \mathrm{d}\mathbf{v} \\
& + \boldsymbol{\lambda}^{(1)} \cdot \int [\nu_{12}(\mathcal{G}_{12} - f_1) m_1 + \nu_{21}(\mathcal{G}_{21} - f_2) m_2] \mathbf{v} \, \mathrm{d}\mathbf{v} \\
& + \lambda^{(2)} \cdot \int [\nu_{12}(\mathcal{G}_{12} - f_1) m_1 + \nu_{21}(\mathcal{G}_{21} - f_2) m_2] |\mathbf{v}|^2 \, \mathrm{d}\mathbf{v} = 0.
\end{aligned}$$

Adding this vanishing term to the dissipation terms yields

$$\begin{aligned}
D_{12} + D_{21} &= D_{12} + D_{21} - I \\
&= \int \nu_{12}(\mathcal{G}_{12} - f_1) \log \frac{f_1}{\mathcal{G}_{12}} \, \mathrm{d}\mathbf{v} + \int \nu_{21}(\mathcal{G}_{21} - f_2) \log \frac{f_2}{\mathcal{G}_{21}} \, \mathrm{d}\mathbf{v} \overset{(2.87)}{\leq} 0
\end{aligned}$$

with equality if and only if $f_1 = \mathcal{G}_{12} = \mathcal{M}[f_1]$ and $f_2 = \mathcal{G}_{21} = \mathcal{M}[f_2]$ using the same arguments as above. Moreover, a direct calculation determines the corresponding mean velocities and temperatures to be

$$\mathbf{u}_1 = \mathbf{u}_2 = -\frac{\boldsymbol{\lambda}^{(1)}}{\lambda^{(2)}} \quad \text{and} \quad T_1 = T_2 = -\frac{1}{2\lambda^{(2)}}.$$

So in this special set-up (equilibrium), the macroscopic quantities $\mathbf{u}_i$ and T_i can be expressed as functions of $\boldsymbol{\lambda}_{ij}$ even if, in general, this is not possible. □

Remark 2.2.8. *The entropy in Definition 2.1.30 is given by an auxiliary funcion* $\tilde{h}(z) = z \log z$ *instead of* $h(z) = z \log z - z$. *But the difference between those only contributes to an additional, constant term (the mass) when integrating with respect to* $\mathbf{v}$. *In the end, the statement of the* $\mathcal{H}$*-Theorem remains the same.*

In order to end the introduction of this model, we specify the velocity-dependent collision frequencies of our interest.

2.2.3 Cross sections and collision frequencies

We give a short introduction into cross sections and collision frequencies in this section. For ease, we start with kernels for one species, but in the end we give the formulas which are used in our numerical simulations for two species.

Cross sections

Cross sections represent the fundamental object regarding the underlying physics in the Boltzmann model, see Definitions 2.1.26, 2.1.27. By a careful handling, the validity of the Boltzmann equation can even be extended, e.g. to dense plasmas [HHM17b, SM16].

Cross sections basically depend on the interaction potential which describes how particles behave near each other. For instance, hard spheres correspond to a collision behavior like billiard balls leading to a constant cross section (2.92). Whereas the Coulomb potential $\phi(r) = \frac{1}{r}$ applies for charged particles, r being the distance of two particles. Then the cross section is given by Rutherford's formula (2.93).

In [Max67], Maxwell derived implicit formulas for the collision kernel in terms of the interaction potential. Only in the two mentioned cases above, they can be made explicit. However, other important interaction potentials are the inverse-power laws

$$\phi(r) = \frac{1}{r^{s-1}}, \quad s > 2.$$

These are often used in physics and modelling and lead to so-called variable-sphere collision kernels. Van der Waals interactions for example typically correspond to $s = 7$ and ion-neutral interactions to $s = 5$ [Vil02]. The problem with this kind of interaction potential is the infinite range which results in a nonintegrable singularity. One can control the collisions with such very large impact parameters ('grazing collisions') by cutting off the integral such that it gets integrable. Regardless of the difficulties, one can show that these collision kernels factor up into a kinetic and an angular part like

$$K(|\mathbf{v}-\mathbf{v}_*|, \boldsymbol{\xi}) = |\mathbf{v}-\mathbf{v}_*| \cdot \sigma(|\mathbf{v}-\mathbf{v}_*|, \theta) = K_k(|\mathbf{v}-\mathbf{v}_*|) \cdot K_a(\cos\theta)$$

with

$$K_k = |\mathbf{v}-\mathbf{v}_*|^\gamma \quad \text{and} \quad K_a(\cos\theta)\sin^{N-2}\theta \sim \text{const} \cdot \theta^{-(1+\nu)}$$
$$\gamma = \frac{s-5}{s-1}, \quad N = 3, \quad \nu = \frac{2}{s-1}.$$

Often, $s > 5$ ($\gamma > 0$) is referred to hard potentials and $s < 5$ ($\gamma < 0$) is referred to soft potentials. The special case $s = 5$ ($\gamma = 0$) leads to a pure dependence of the collision kernel on the angle θ. The latter theoretical model corresponds to so-called *Maxwellian molecules*.

It is not clear which range of values for s should be considered. The range can amount to $s \in (2, \infty)$, where $s = 2$ is the limit for Coulomb interaction which at first glance does not really fit to Boltzmann, see Remark 2.2.9.

Just by this short introduction, it is evident that much effort in modelling can be spent. Especially for the comparison with physics, a careful determination of constants is needed. In the following, we list the most common cross sections used in the literature. More information can be found in [Bir94, KM91, Bir80, Vil02].

In the field of aerospace engineering, the *variable-soft-spheres* cross section is often used. It takes the form

$$\sigma(|\mathbf{v}-\mathbf{v}_*|,\theta) = \frac{\alpha}{2}\frac{\cos^{2\alpha-1}(\theta/2)\sin(\theta/2)}{\sin(\theta)}\frac{d_{\text{ref}}^2}{\Gamma(\frac{5}{2}-\omega)}\left(\frac{2k_B T_{\text{ref}}}{\mu|\mathbf{v}-\mathbf{v}_*|^2}\right)^{\omega-\frac{1}{2}}, \tag{2.88}$$

where μ is the reduced mass and $\alpha, T_{\text{ref}}, d_{\text{ref}}$ and $\omega = \frac{\gamma}{2}$ are reference parameter for the material in question. However, in other fields, this formula is used less.

Taking $\alpha = 1$ in (2.88) leads to the *variable-hard-spheres* cross section

$$\sigma(|\mathbf{v}-\mathbf{v}_*|) = \frac{d_{\text{ref}}^2}{4\Gamma(\frac{5}{2}-\omega)}\left(\frac{2k_B T_{\text{ref}}}{\mu|\mathbf{v}-\mathbf{v}_*|^2}\right)^{\omega-\frac{1}{2}}. \tag{2.89}$$

A very interesting cross section for Boltzmann and BGK models is when the collision kernel $K = \sigma \cdot |\mathbf{v}-\mathbf{v}_*|$ has no velocity dependence. This is the case for Maxwellian molecules, setting $\omega = 1$ in (2.88):

$$\sigma(|\mathbf{v}-\mathbf{v}_*|,\theta) = \frac{\alpha}{2}\frac{\cos^{2\alpha-1}(\theta/2)\sin(\theta/2)}{\sin(\theta)}\frac{2d_{\text{ref}}^2}{\pi^{\frac{1}{2}}|\mathbf{v}-\mathbf{v}_*|}\left(\frac{2k_B T_{\text{ref}}}{\mu}\right)^{\frac{1}{2}}. \tag{2.90}$$

More frequently, the *pseudo-Maxwell molecules* are incorporated. Here, the angular dependence in (2.90) is overcome by additionally having $\alpha = 1$:

$$\sigma(|\mathbf{v}-\mathbf{v}_*|) = \frac{d_{\text{ref}}^2}{2\pi^{\frac{1}{2}}|\mathbf{v}-\mathbf{v}_*|}\left(\frac{2k_B T_{\text{ref}}}{\mu}\right)^{\frac{1}{2}}, \tag{2.91}$$

which also results in a constant in $\mathbf{v}$ collision rate for BGK models.

Eventually, for $\alpha = 1$ and $\omega = \frac{1}{2}$ we obtain the constant *hard-spheres* cross section

$$\sigma = \frac{d_{\text{ref}}^2}{4}. \tag{2.92}$$

For plasmas, a distinct kind of cross sections is used. In this state of matter, we expect a velocity dependence which scales like $|\mathbf{v}-\mathbf{v}_*|^4$. However, corresponding choices for α and ω in (2.88) led to problems, e.g. a $\frac{1}{\Gamma(0)}$ term. Nevertheless, in plasmas Coulomb collisions are typically assumed. For these, we can apply the *Rutherford* cross section

$$\sigma(|\mathbf{v}-\mathbf{v}_*|,\theta) = \left(\frac{Z^2e^2}{2m|\mathbf{v}-\mathbf{v}_*|^2\sin^2(\theta/2)}\right)^2. \tag{2.93}$$

A derivation of (2.93) can be found e.g. in [LL69].

Remark 2.2.9 (Boltzmann operator and charged particles). *Plugging in the Rutherford cross section* (2.93) *into the Boltzmann collision operator* (2.26) *leads to a diverging integral [Vil02, HHM17a]. This is because of the slow decay of the Coulomb potential* $\phi(r) = \frac{1}{r}$ *and the corresponding singularity in the cross section for small-angle collisions. A standard remedy is to establish a screening leading to the Debye potential*

$$\phi(r) = \frac{e^{-r/\lambda_D}}{r}$$

or to introduce a cut-off [SM16]. Another work-around is to approximate the Boltzmann operator. A small-angle approximation yields the Landau or Fokker-Planck collision operator. But also a BGK collision operator can be used with a corresponding handling of the collision frequency what is discussed next.

Collision frequencies

Cross sections are part of the Boltzmann model, and these are inherited to the BGK models because the collision frequencies can be seen as averaged cross sections. Now, even more modelling comes into play since issues of complexity force us to establish simplifications. Especially integrability issues occur: In the Boltzmann model, an 'infinite cancellation' between gain and loss terms takes place. The BGK derivation naturally splits these gain and loss terms such that we have to deal with singularities. Often, these non-trivial calculations are done for specific cross sections. For example in [HHM17a], they perform detailed computations in order to achieve their (velocity-independent) collision frequencies.

In Section 2.2.1, we already motivated why we insist on a collision frequency which does depend on the relative velocity. Before we come to a derivation of closed formulas for collision frequencies in a plasma, we want to clarify our expectations. On the one hand, for small relative velocities we expect a considerable impact of interactions. On the other hand, for large relative velocities we expect that the influence of the interactions decreases, see Figure 2.8. In Figure 2.9, we show that our final formula matches these expectations.

Derivation The most obvious way to derive a collision frequency is to use its definition (2.48)

$$\nu(\mathbf{x}, \mathbf{v}, t) = \int_{\mathbb{R}^3} \int_{S^2} \mathcal{M}(\mathbf{x}, \mathbf{v}_*, t) |\mathbf{v} - \mathbf{v}_*| \sigma(|\mathbf{v} - \mathbf{v}_*|, \xi) \, \mathrm{d}\xi \, \mathrm{d}\mathbf{v}_*$$

which comes intrinsically by the derivation of the BGK model [Str97, HHM17a]. This is used for Maxwell molecules due to their special cross section (2.90). But in a plasma, the Rutherford cross section scales like

$$\sigma \sim \frac{1}{|\mathbf{v} - \mathbf{v}_*|^4 \sin^4(\theta/2)}.$$

Unfortunately, this has a singularity to be dealt with. Often, this is handled by an angular cut-off. However, the integral still blows up near the cut-off angle for charged particles. This is related to the grazing collisions limit in the Boltzmann operator [GH14] which does not seem to be conducive for BGK terms.

In order to reduce the complexity, one often considers the *momentum transfer* cross section

$$\sigma_{\mathrm{MT}}(|\mathbf{v}-\mathbf{v}_*|) = 2\pi \int_0^\pi \sigma(|\mathbf{v}-\mathbf{v}_*|,\theta)\sin^2(\theta/2)\sin\theta\,\mathrm{d}\theta.$$

This is one of the fundamental quantities in the theory of binary collisions from a physical point of view [Vil02, SJB66, DB94]. Inserting the Rutherford cross section (2.93), we obtain

$$\begin{aligned}\sigma_{\mathrm{MT}}(|\mathbf{v}-\mathbf{v}_*|) &= 4\pi\frac{(Z^2e^2)^2}{(2m|\mathbf{v}-\mathbf{v}_*|^2)^2}\int_0^\pi \frac{1}{\sin^4(\theta/2)}\sin^2(\theta/2)\sin\theta\,\mathrm{d}\theta\\ &= 8\pi\frac{(Z^2e^2)^2}{(2m|\mathbf{v}-\mathbf{v}_*|^2)^2}\int_0^\pi \frac{\cos(\theta/2)}{\sin(\theta/2)}\,\mathrm{d}\theta\\ &= 16\pi\frac{(Z^2e^2)^2}{(2m|\mathbf{v}-\mathbf{v}_*|^2)^2}\log(\sin(\theta/2))\Big|_0^\pi\end{aligned}$$

by using trigonometric identities. Again, we run into a problem concerning a singularity. Therefore, we perform an angular cut-off using an argument related to Debye screening [SM16]:

$$\begin{aligned}\sigma_{\mathrm{MT}}(|\mathbf{v}-\mathbf{v}_*|) &\approx 16\pi\frac{(Z^2e^2)^2}{(2m|\mathbf{v}-\mathbf{v}_*|^2)^2}\log(\sin(\theta/2))\Big|_{\theta_{\min}}^\pi\\ &= -16\pi\frac{(Z^2e^2)^2}{(2m|\mathbf{v}-\mathbf{v}_*|^2)^2}\log\left(\sin\frac{\theta_{\min}}{2}\right)\\ &:= 16\pi\frac{(Z^2e^2)^2}{(2m|\mathbf{v}-\mathbf{v}_*|)^2)^2}\log\Lambda.\end{aligned}\tag{2.94}$$

The so-called *Coulomb logarithm* $\log\Lambda$ plays another important role in modelling and is discussed in Remark 2.2.10. We plug in the momentum transfer cross section (2.94) into (2.48)

$$\nu(\mathbf{v}) = \int_{\mathbb{R}^3}\int_{S^2}\mathcal{M}(\mathbf{v}_*)\sigma_{\mathrm{MT}}(|\mathbf{v}-\mathbf{v}_*|)|\mathbf{v}-\mathbf{v}_*|\,\mathrm{d}\xi\,\mathrm{d}\mathbf{v}_*.$$

This gives us another singularity for $|\mathbf{v}-\mathbf{v}_*|=0$ which cannot be overcome easily. Therefore, we follow another approach.

The mean free time between collisions scales like $\frac{1}{n|\mathbf{v}|\sigma}$ [LM84, Sch06]. As the collision frequency represents the number of collisions per time unit, we follow the approach of the ad-hoc collision rate

$$\nu(\mathbf{v}) = n|\mathbf{v}|\sigma(\mathbf{v}).$$

Several choices need to be made. We conveniently take [LM84]

$$\begin{aligned}\nu(\mathbf{v}) &= n|\mathbf{v}-\mathbf{u}|\sigma_{\mathrm{MT}}(|\mathbf{v}-\mathbf{u}|) \\ &\approx 4\pi n \frac{(Z^2e^2)^2}{m^2} \log\Lambda \cdot \frac{1}{\delta + |\mathbf{v}-\mathbf{u}|^3}\end{aligned}$$

where we use the relative velocity $|\mathbf{v}-\mathbf{u}|$ at the kinetic level. In order to avoid a singularity at $|\mathbf{v}-\mathbf{u}| = 0$, we have added some $\delta > 0$ in the denominator. This is specified for our numerical simulations in Section 5.3, depending on the velocity grid.

In the following remark, we discuss the Coulomb logarithm.

Remark 2.2.10 (Coulomb logarithm). *The Boltzmann equation is not capable to model charged particles for which a Coulomb potential applies, see Remark 2.2.9. In 1937, Landau [Lan37] approximated the Boltzmann operator by a small angle scattering which works for the Coulomb interactions and led to the Fokker-Planck-Landau equation [Lan37, RMJ57]. It holds*

$$\mathscr{Q}_{\mathrm{Boltz}} \approx \log\Lambda \cdot \mathscr{Q}_{\mathrm{FPL}} + O(1)$$

where $\log\Lambda$ *is the Coulomb logarithm and FPL stands for Fokker-Planck-Landau.*

The Coulomb logarithm encounters minimal and maximal impact parameters arising through truncating corresponding limits in an integral [SM16]. A simple Coulomb logarithm, appropriate for classical plasmas reads

$$\log\Lambda = \log\left(\frac{\lambda_D}{b_{90}}\right), \tag{2.95}$$

where λ_D *is the Debye length and* b_{90} *is the classical distance of closest approach (and the* 90° *deflection radius, respectively). These depend on the moments of the system (density and temperature). The argument of the logarithm gets small for little collisions. To stay positive, we use*

$$\log\Lambda = \frac{1}{2}\log\left(1 + \frac{\lambda_D^2}{b_{90}^2}\right). \tag{2.96}$$

More sophisticated approaches can be followed. Instead of the momentum transfer cross section (2.94), one might use more elaborated ones as can be found in [SM16]. But as we already mentioned earlier, much effort in modelling can be spent here. For our numerical simulations we are happy with the following collision frequencies.

Velocity-dependent collision frequency formulas To conclude, we summarize and specify the formulas we use in our numerical simulations. All formulae are given in CGS units, see Appendix D. This also requires the consistent use of Boltzmann's constant k_B. In all other parts of this thesis, we suppress k_B for ease of presentation.

We make the following assumption on the collision frequencies $\nu_{ij}(\mathbf{x}, \mathbf{v}, t)$.

Assumption 2.2.11. *The space and time dependence of the collision frequencies ν_{ij} arise via a dependence on the mass densities $\rho_i(\mathbf{x}, t)$, the mixture mean velocity $\mathbf{u}_{\text{mix}}(\mathbf{x}, t)$ and the mixture temperature $T_{\text{mix}}(\mathbf{x}, t)$. Furthermore, because the collisional process conserves these quantities, the collision frequencies ν_{ij} are independent of time in the space homogeneous setting.*

This relationship between collision frequencies and moments is typical for standard collision rates in the literature and follows from cross section definitions; see for example [KT73, HHM17a].

We use the following velocity-dependent collision frequencies

$$\nu_{ij}(\mathbf{v}) = n_j 4\pi \left(\frac{Z_i Z_j e^2}{2\mu_{ij}}\right)^2 \frac{1}{\delta_{ij} + |\mathbf{v} - \mathbf{u}_{\text{mix}}|^3} \log \Lambda_{ij} \tag{2.97}$$

where $\mu_{ij} = \frac{m_i m_j}{m_i + m_j}$ is the reduced mass; $Z_i e$, $Z_j e$ are the charges of the two particles; and $\log \Lambda_{ij}$ is the Coulomb logarithm given by

$$\log \Lambda_{ij} = \frac{1}{2} \log \left(1 + \frac{\lambda_D^2}{b_{90,ij}^2}\right). \tag{2.98}$$

The regularization parameter δ_{ij} will depend on the corresponding velocity grid in order to involve an adequate scaling. We set the classical distance of closest approach to be

$$b_{90,ij} = \frac{Z_i Z_j e^2}{T_{\text{mix}}}. \tag{2.99}$$

For the Debye length λ_D we use the following formulae

$$\lambda_D = \left(\frac{1}{\lambda_e^2} + \frac{1}{\lambda_I^2}\right)^{-1/2}$$

with

$$\lambda_e = \left(\frac{T_{\text{mix}}}{4\pi n_e e^2}\right)^{1/2} \quad \text{and} \quad \lambda_I = \left(\sum_{i \in S} \frac{1}{\lambda_i^2}\right)^{-1/2},$$

where

$$\lambda_i = \left(\frac{T_{\text{mix}}}{4\pi n_i Z_i^2 e^2}\right)^{1/2} \quad \text{and} \quad n_e = \sum_{i \in S} Z_i n_i,$$

with the Boltzmann constant k_B. The mixture quantities $\mathbf{u}_{\text{mix}}$ and T_{mix} defined in (2.8) and (2.9) are inserted into these formulas to determine the collision frequencies used in the model. There are modifications that can be made to ensure that these formulas (e.g. screening length) apply to a wider range of plasma regimes, but this should be a reasonable start.

The shape of (2.97) is shown in Figure 2.9.

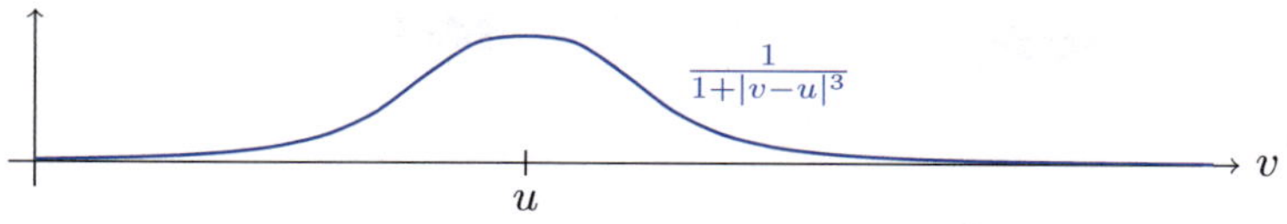

Figure 2.9: The shape of the velocity-dependent collision frequency (2.97) is illustrated.

Velocity-independent collision frequency formulas For a comparison, we use several velocity-independent collision frequencies of comparable size where we guarantee that $\frac{\nu_{12}}{\nu_{21}} = \frac{n_2}{n_1}$. A convenient choice in plasma physics [SM16] is to take

$$\tilde{\nu}_{ij} = \rho_j 4\pi \left(\frac{Z_i Z_j e^2}{2\mu_{ij}} \right)^2 \frac{1}{\delta_{ij} + v_T^3} \log \Lambda_{ij}, \tag{2.100}$$

where the deviation from the mean velocity $|\mathbf{v} - \mathbf{u}_{\text{mix}}|$ is approximated by the thermal velocity $v_T = \sqrt{k_B T_{\text{mix}}/(2\mu_{ij})}$. However, this choice seems arbitrary. When replacing $|\mathbf{v} - \mathbf{u}_{\text{mix}}|$ by v_T, the observation

$$\int |\mathbf{v} - \mathbf{u}|^2 f \, \mathrm{d}\mathbf{v} = 3n\frac{T}{m} = 3nv_T^2 \tag{2.101}$$

allows also for replacing $|\mathbf{v} - \mathbf{u}_{\text{mix}}|$ by $\sqrt{3}v_T$. That is why we need to normalize a corresponding constant collision frequency in some way. In order to calibrate possible constants we average the velocities by

$$\hat{v}^3 = \frac{\int |\mathbf{v} - \mathbf{u}_{\text{mix}}|^3 \mathcal{M}_{ij}(\mathbf{v}) \, \mathrm{d}\mathbf{v}}{\int \mathcal{M}_{ij}(\mathbf{v}) \, \mathrm{d}\mathbf{v}} \tag{2.102}$$

where

$$\mathcal{M}_{ij}(\mathbf{v}) = \mathcal{M}[n_i, \mathbf{u}_{\text{mix}}, T_{\text{mix}}, 2\mu_{ij}]. \tag{2.103}$$

Replacing v_T^3 in (2.100) by (2.102), gives us a second option for a corresponding velocity-independent collision frequencies $\hat{\nu}_{ij}$ resulting in

$$\hat{\nu}_{ij} = \rho_j 4\pi \left(\frac{Z_i Z_j e^2}{2\mu_{ij}} \right)^2 \frac{1}{\delta_{ij} + \hat{v}^3} \log \Lambda_{ij}. \tag{2.104}$$

Another natural way to obtain velocity-independent collision frequencies is to average $\nu_{ij}(\mathbf{v})$ given by (2.97) themselves leading to

$$\bar{\nu}_{ij} = \frac{\int \nu_{ij}(\mathbf{v}) \mathcal{M}_{ij}(\mathbf{v}) \, \mathrm{d}\mathbf{v}}{\int \mathcal{M}_{ij}(\mathbf{v}) \, \mathrm{d}\mathbf{v}} \tag{2.105}$$

with the Maxwellians $\mathcal{M}_{ij}(\mathbf{v}) = \mathcal{M}[n_i, \mathbf{u}_{\text{mix}}, T_{\text{mix}}, 2\mu_{ij}]$.

For later reference, we also give the velocity-independent collision frequencies derived in [HHM17a]. We start with collision frequencies describing ion-ion interactions in dense plasma. The collision frequencies ν_{ij}^M in (2.106) and (2.109) ensure that the relaxation rates of the momenta for the model in [HHM17a] coincide with the one of the multi-species Boltzmann equations. And the collision frequencies ν_{ij}^T in (2.107) and (2.110) ensure the corresponding relaxation rates for the temperatures. It is

$$\nu_{ij}^M = \frac{128}{3} \frac{\pi^2}{(2\pi)^{3/2}} \frac{n_j}{m_i} (Z_j Z_i e^2)^2 \frac{\sqrt{m_i m_j}(m_i + m_j)}{(m_i T_j + m_j T_i)^{3/2}} \Psi(\gamma_{ij}) \tag{2.106}$$

$$\nu_{ij}^T = \frac{256}{3} \frac{\pi^2}{(2\pi)^{3/2}} n_j (Z_j Z_i e^2)^2 \frac{\sqrt{m_i m_j}}{(m_i T_j + m_j T_i)^{3/2}} \Psi(\gamma_{ij}) \tag{2.107}$$

with the charges $Z_1 e, Z_2 e$, the elementary charge e, the function

$$\Psi(x) = \begin{cases} -\frac{1}{4}\log\left(\sum_{n=1}^{5} a_n x^n\right) & \text{for} \quad x < 1, \\ \frac{b_0 + b_1 \log x + b_2 \log^2 x}{1 + b_3 x + b_4 x^2} & \text{for} \quad x > 1, \end{cases}$$

where the coefficients a_n, b_n are defined in [SM16]

n	0	1	2	3	4	5
a_n		1.4660	−1.7836	1.4313	−0.55833	0.061162
b_n	0.081033	−0.091336	0.051760	−0.50026	0.17044	

with the short form

$$\gamma_{ij} = \frac{Z_j Z_i e^2}{\lambda} \frac{m_i + m_j}{m_i T_j + m_j T_i}$$

and the screening length

$$\lambda = \left[\frac{4\pi e^2 n_e}{\sqrt{T_{\text{mix}}^2 + \left(\frac{\hbar^2}{3m_e}\right)^2 (3\pi^2 n_e)^{4/3}}} + \sum_{i=1}^{2} \frac{4\pi (Z_i e)^2 \rho_i}{T_{\text{mix}} + 3 Z_i^2 e^2 \left(\frac{4\pi}{3}\frac{\sigma}{Z_i}\right)^{-1/3}} \right]^{-\frac{1}{2}}. \tag{2.108}$$

Here, we use the electron mass m_e, its density $n_e = Z_1 n_1 + Z_2 n_2$ and the mixture temperature T_{mix}, the total charge density (devided by e) $\sigma = \sum Z_i \rho_i$ and the reduced Planck constant $\hbar$.

Collision frequencies describing screened Coulomb interactions with a Coulomb logarithm are given by

$$\nu_{ij}^M = \frac{8}{3}\sqrt{2\pi}\, n_j \sqrt{\frac{m_j}{m_i}} \frac{m_i + m_j}{(m_i T_j + m_j T_i)^{3/2}} (Z_i Z_j e^2)^2 \log \Lambda_{\text{GMS6}}, \tag{2.109}$$

$$\nu_{ij}^T = \frac{16}{3}\sqrt{2\pi}\, n_j \frac{\sqrt{m_i m_j}}{(m_i T_j + m_j T_i)^{3/2}} (Z_i Z_j e^2)^2 \log \Lambda_{\text{GMS6}}. \tag{2.110}$$

The Coulomb logarithm $\log \Lambda_{\mathrm{GMS6}}$ takes the following form

$$\log \Lambda_{\mathrm{GMS6}} = \frac{1}{2} \log \left(1 + \frac{\lambda^2 + a_i^2}{\lambda_{\mathrm{dB},i}^2 + b_{ij}^2} \right) \tag{2.111}$$

considering the screening length (2.108), the ion sphere radius $a_i = \left(\frac{3}{4}\pi \frac{|Z_i|}{\sigma} \right)^{1/3}$, the deBroglie wave length $\lambda_{\mathrm{dB,i}} = \frac{2\pi\hbar}{\sqrt{m_i T_i}}$ and the distance of closest approach $b_{ij} = \max(0, b_{90,ij})$ using $b_{90,ij}$ from (2.99).

Now, we are equipped with an array of different collision frequencies and we make use of them in Section 5.3. For a moment, we can put them aside. In the following section, we handle collision frequencies with simplicity, but we include quantum effects instead.

2.3 Quantum multi-species BGK models

Classical physics is based on Newton's laws (Axioms 2.1.2–2.1.4). However, at the end of the 19th century, physicists experienced phenomena which could not be explained by those, more severe, which even contradicted them. A new theory was developed at the beginning of the 20th century concerning such effects at very small scales, i.e. atomic scales: quantum mechanics.

Although many phenomena in our daily life can only be understood with the help of quantum physics, it is very challenging to deal with this theory. Even the Nobel laureate Richard Feynman stated: 'I think I can safely say that nobody really understands quantum mechanics.'

2.3.1 Fundamentals

In the following section, we do not go into details of quantum theory and restrict ourselves to some selected concepts needed for the quantum description of kinetic models. For more information, we refer to [Sch07].

Introduction to quantum mechanics

In classical mechanics, a system of particles with masses m_i is fully determined by their positions $\mathbf{x}_i$ and velocities $\mathbf{v}_i = \frac{\mathbf{p}_i}{m_i}$, where $\mathbf{p}_i$ is the corresponding momentum. For quantum particles, we run into several problems.

Firstly, we should track the particles' momenta instead of the velocities. This is due to the fact that e.g. the mass of a photon is zero, but a photon's momentum needs not to vanish. Hence, the notion 'velocity' needs to be treated with care. In this thesis, we only consider nonrelativistic particles with nonvanishing mass such that we do have the relationship $\mathbf{v}_i = \frac{\mathbf{p}_i}{m_i}$. Nevertheless, in quantum theory the momentum variable is requested being the canonic conjugate quantity to the space variable. So the phase space is now given by $(\mathbf{x}, \mathbf{p})$.

A second issue is the *Heisenberg uncertainty principle*. We do not go into details here, but we want to mention that it is possible to exactly measure either position or momentum of a particle, but not both. This means that a system cannot be fully determined, instead a quantum system is always described with uncertainties and probabilities, respectively.

These uncertainties lead to another concern, namely the fact that quantum particles can be *identical*. This means that they cannot be distinguished, not even in principle. To see this, imagine two electrons. Their intrinsic physical properties are all the same, such as mass, charge and spin. Hence, distinguishing these electrons would only be possible by tracking the trajectory of each particle. Due to the Heisenberg uncertainty principle, the trajectory cannot be determined with infinite precision which makes an exact tracking impossible: The electrons are *indistinguishable*.

Actually, the further behavior of identical particles leads to a classification. A subset of quantum particles follows *Pauli's exclusion principle* which states that any quantum state within a quantum system can only be occupied by at most one particle.

Definition 2.3.1 (Fermions and bosons). *A fermion is a quantum particle obeying Pauli's exclusion principle. It is also classified by possessing a half-integer spin.*

A quantum particle whose spin quantum number has an integer value is called boson. Additionally, it does not follow Pauli's exclusion principle.

Electrons and protons are fermions. So in the case of electrons in a poly-electron atom, the Pauli exclusion principle has the following consequence. If two electrons are located in the same orbital, three out of the four quantum numbers are the same. Thus, these electrons must have a different spin quantum number ($\pm\frac{1}{2}$). So this orbital is full, and additional electrons must reside in different orbitals.

Bosons, such as the Higgs boson or photons, behave differently. One quantum state can be occupied by any number of identical bosons which is the case e.g. for lasers or Bose-Einstein-condensates. For formal reasons, we exclude photons in this thesis.

We end this short introduction into quantum mechanics and continue by carrying quantum effects into kinetic theory.

Quantum Boltzmann equation

Around 1930, the physicists Nordheim, Ueling and Uhlenbeck developed a kinetic model for a quantum gas [FN28, KN30, UU33, Ueh34]. They use a kinetic description as for classical gases, but they take quantum effects of the particles into account.

To be precise, we introduce the quantum distribution function.

Definition 2.3.2 (Quantum distribution function). *A function $f : \mathbb{R}^3 \times \mathbb{R}^3 \times \mathbb{R}_0^+ \to \mathbb{R}_0^+$ is called a distribution function if and only if $f(\mathbf{x}, \mathrm{d}\mathbf{p}, t)\, \mathrm{d}\mathbf{x}\, \mathrm{d}\mathbf{p}$ is the number of particles with momenta in $(\mathbf{p}, \mathbf{p} + \mathrm{d}\mathbf{p})$ located at $(\mathbf{x}, \mathbf{x} + \mathrm{d}\mathbf{x})$ at time t.*

For fermions, we additionally require

$$f(\mathbf{x}, \mathbf{p}, t) < 1. \tag{2.112}$$

The upper bound for fermions is physically motivated as described in the following remark.

Remark 2.3.3 (Bounds for fermions). *The additional upper bound for distribution functions of fermions comes by integrability conditions. In general, we require that all 'physical' quantities are bounded, i.e.*

$$\int f(1+|\mathbf{p}|^2)\,\mathrm{d}\mathbf{p} < \infty \quad \textit{and} \quad \mathcal{H}[f] < \infty,$$

where $\mathcal{H}$ denotes the entropy, see Definition 2.3.7. These requirements represent integrability conditions and for fermions additionally lead to (2.112). *[EMV03]*

Now, we can formulate the quantum Boltzmann equation.

Definition 2.3.4 (Quantum Boltzmann equation). *The quantum Boltzmann equation for a distribution function $f = f(\mathbf{x}, \mathbf{p}, t)$ reads*

$$\partial_t f + \mathbf{v} \cdot \nabla_{\mathbf{x}} f = \mathcal{Q}[f] \tag{2.113}$$

with the collision operator

$$\mathcal{Q}[f](\mathbf{x},\mathbf{p},t) = \\ \int_{\mathbb{R}^3}\int_{\mathcal{S}^2} (f' f'_*(1-\tau f)(1-\tau f_*) - f f_*(1-\tau f')(1-\tau f'_*)) K(|\mathbf{p}-\mathbf{p}_*|,\xi)\,\mathrm{d}\xi\,\mathrm{d}\mathbf{p}_*, \tag{2.114}$$

where $f = f(\mathbf{p})$, $f' = f(\mathbf{p}')$, $f_* = f(\mathbf{p}_*)$, $f'_* = f(\mathbf{p}'_*)$; $K(|\mathbf{p}-\mathbf{p}_*|,\xi)$ is the collision kernel, ξ represents the unit vector in scattering direction and $\mathbf{v} = \frac{\mathbf{p}}{m}$.

This equation is often called Ueling-Uhlenbeck equation or Nordheim equation in the literature.

The collision operator $\mathcal{Q}$ describes the effect of interactions between particles. The pre-collisional and post-collisional momenta follow the conservation of momentum and energy at the particle level as in the classical case, see Theorems 2.1.6 and 2.1.7; that is

$$\mathbf{p} + \mathbf{p}_* = \mathbf{p}' + \mathbf{p}'_*$$
$$|\mathbf{p}|^2 + |\mathbf{p}_*|^2 = |\mathbf{p}'|^2 + |\mathbf{p}'_*|^2.$$

In contrast to the classical Boltzmann operator (2.26), the operator includes additional terms in the quantum case (2.114) taking into accout the degeneracy of quantum particles. For fermions $\tau = +1$, these guarantee that Pauli's principle is satisfied, and for bosons $\tau = -1$, the possibility to cluster is given.[2]

The notions mass, momentum and energy can be carried over from the classical theory. For clearness regarding variables, we provide the following definition. For clearness regarding units, we refer to Appendix D.2.

[2] In the literature, the sign of τ is not used consistently. When τ is plugged in, the final equations matter.

Definition 2.3.5 (Macroscopic quantities in the quantum case). *Let f be the distribution function of a quantum gas with particle mass m. We denote the species' mass density by*

$$N = \int_{\mathbb{R}^3} f \, \mathrm{d}\mathbf{p},$$

the species' momentum by

$$\mathbf{P} = \int_{\mathbb{R}^3} f \mathbf{p} \, \mathrm{d}\mathbf{p}$$

and the species' energy density by

$$E = \int_{\mathbb{R}^3} f \frac{|\mathbf{p}|^2}{2m} \, \mathrm{d}\mathbf{p}.$$

We define the kinetic temperature by

$$T = \frac{2}{3N} E - \frac{1}{3m} \frac{|\mathbf{P}|^2}{N^2} = \frac{1}{3mN} \int_{\mathbb{R}^3} f \left| \mathbf{p} - \frac{\mathbf{P}}{N} \right|^2 \mathrm{d}\mathbf{p}. \tag{2.115}$$

The notion of temperature seems intuitively clear. However, it is a concept for equilibrium states only. We postpone a further discussion of this issue to Remark 2.3.10.

The quantum Boltzmann equation fulfils the following invariance property.

Theorem 2.3.6 (Conservation properties for the quantum Boltzmann equation). *The quantum Boltzmann equation conserves mass, momentum and energy.*

Proof. A proof for a more general multi-species quantum Boltzmann equation is provided e.g. in [BKPY21]. □

The trend to equilibrium is of central interest. Therefore, the entropy needs to be investigated.

Definition 2.3.7 (Quantum entropy). *The entropy density functional of a quantum gas is defined by*

$$\mathcal{H}[f] = \int [f \log f + \tau(1 - \tau f) \log(1 - \tau f)] \, \mathrm{d}\mathbf{p}. \tag{2.116}$$

Entropy minimizers are the equilibrium states we are looking for. We already know that the distribution function of a classic gas converges to a Maxwellian distribution for $t \to \infty$. The analogues of quantum particles are called Fermi-Dirac distribution functions for fermions and Bose-Einstein distribution functions for bosons. We introduce these distribution functions in the following definition and give the formal statement below.

Definition 2.3.8 (Specific quantum distribution functions). *A Maxwellian $\mathcal{M}(\mathbf{x}, \mathbf{p}, t) \geq 0$ is a distribution function of the specific form*

$$\mathcal{M}(\mathbf{x}, \mathbf{p}, t) = \mathcal{M}[a, \mathbf{b}, c, m](\mathbf{x}, \mathbf{p}, t) = \frac{1}{e^{ma\left|\frac{\mathbf{p}}{m}-\mathbf{b}\right|^2+c}} \tag{2.117}$$

with $a > 0$. A Fermi-Dirac distribution function $\mathcal{F}(\mathbf{x}, \mathbf{p}, t) \geq 0$ is defined by

$$\mathcal{F}(\mathbf{x}, \mathbf{p}, t) = \mathcal{F}[a, \mathbf{b}, c, m](\mathbf{x}, \mathbf{p}, t) = \frac{1}{e^{ma\left|\frac{\mathbf{p}}{m}-\mathbf{b}\right|^2+c}+1}. \tag{2.118}$$

In contrast to a Bose-Einstein distribution function $\mathcal{B}(\mathbf{x}, \mathbf{p}, t)$ being given by

$$\mathcal{B}(\mathbf{x}, \mathbf{p}, t) = \mathcal{B}[a, \mathbf{b}, c, m](\mathbf{x}, \mathbf{p}, t) = \frac{1}{e^{ma\left|\frac{\mathbf{p}}{m}-\mathbf{b}\right|^2+c}-1} \tag{2.119}$$

with

$$ma\left|\frac{\mathbf{p}}{m}-\mathbf{b}\right|^2+c < 0.$$

These three distribution functions can be summarized in the general formulation

$$\mathcal{E}_\tau(\mathbf{x}, \mathbf{p}, t) = \mathcal{E}_\tau[a, \mathbf{b}, c, m](\mathbf{x}, \mathbf{p}, t) = \frac{1}{e^{ma\left|\frac{\mathbf{p}}{m}-\mathbf{b}\right|^2+c}+\tau} = \begin{cases} \mathcal{M} & \text{for } \tau = 0, \\ \mathcal{F} & \text{for } \tau = +1, \\ \mathcal{B} & \text{for } \tau = -1. \end{cases} \tag{2.120}$$

Theorem 2.3.9 ($\mathcal{H}$-Theorem for the quantum Boltzmann equation). *The entropy is dissipated in the space homogeneous quantum gas*

$$\partial_t \mathcal{H} \leq 0$$

with equality if and only if

$$f = \mathcal{E}_\tau.$$

Proof. A proof can be found e.g. in [EMV03]. □

For classic particles, the parameters $a, \mathbf{b}, c$ in the local equilibrium (Maxwellian) are related explicitly to the macroscopic quantities $N, \mathbf{P}, T$. (This is an easy calculation.) But for general quantum particles, it is more complicated. We only have a relationship between $\mathbf{b}$ and $\mathbf{P}$ (see the proof of Theorem 2.3.13), and a can be related to the concept of physical temperature (see Remark 2.3.10).

Remark 2.3.10 (The concept of temperature). *The physical temperature ϑ is uniquely defined for equilibrium states by the second law of thermodynamics, see Axiom 2.1.21 and (2.13). It can be seen as the price at which a system offers energy in return for entropy [Hin20].*

In kinetic theory, this concept is extended for nonequilibrium states, and we formally introduce the (kinetic) temperature T in (2.6)/(2.115). For monatomic ideal and classic gases, it holds $T = \vartheta$. Hence, the temperature (physical and kinetic) can be seen as a measure for the internal energy. But in more general settings (e.g. discrete-velocity or quantum models), T does not equal the physical temperature ϑ even though the macroscopic quantities like density, pressure and energy still coincide with the corresponding moments of the distribution function. [Cer97]

In other words: The kinetic temperature is closely related to the total energy minus the kinetic energy. However, the physical temperature prices the internal energy. Only in special cases, this 'price' can be reduced to a relationship between the total energy and the translational motion only. In general, the contribution of heat, pressure, volume,... to the total energy cannot be expressed by the kinetic temperature.

We yet 'find' the physical temperature in our quantum formulation by the following observation. In thermodynamic theory, the physical temperature appears in the equilibrium canonical distribution. The latter one is proportional to $\exp(-\beta E)$ where E is the total energy of the gas and β is the inverse temperature: $\beta^{-1} = k_B \vartheta$ with the Boltzmann constant k_B. With this fact in mind, we look at the local equilibria $\mathcal{E}_\tau$ and relate the physical temperature to the equilibrium coefficient a.

Additionally, we can formally relate the notions of temperatures using Bose-Einstein and Fermi-Dirac functions, respectively. [HJ11]

The quantum Boltzmann operator (2.114) is very complex making practical applications challenging, and there are many interesting questions [Vil02, EMV03]. In order to reduce complexity, quantum BGK models are widely used. In the following section, we present a consistent multi-species quantum BGK model.

2.3.2 A consistent multi-species quantum BGK model

Recently, a consistent multi-species quantum BGK model was published in [BKPY21]. We recall this model for two species, but everything can also be extended to more species because only binary interactions are taken into account. We consider two distribution functions $f_1 = f_1(\mathbf{x}, \mathbf{p}, t) \geq 0$ and $f_2 = f_2(\mathbf{x}, \mathbf{p}, t) \geq 0$ for species with masses m_1 and m_2, respectively, with the phase space variables position $\mathbf{x} \in \mathbb{R}^3$ and momentum $\mathbf{p} \in \mathbb{R}^3$ and time $t \geq 0$.

Definition 2.3.11 (Quantum multi-species BGK equations). *The quantum multi-species BGK equations in [BKPY21] read*

$$\begin{aligned} \partial_t f_1 + \frac{\mathbf{p}}{m_1} \cdot \nabla_{\mathbf{x}} f_1 &= \mathcal{Q}_1[\{f_1, f_2\}], \\ \partial_t f_2 + \frac{\mathbf{p}}{m_2} \cdot \nabla_{\mathbf{x}} f_2 &= \mathcal{Q}_2[\{f_1, f_2\}] \end{aligned} \tag{2.121}$$

with the collision operators

$$\mathcal{Q}_1[\{f_1, f_2\}] = \mathcal{Q}_{11}[f_1, f_1] + \mathcal{Q}_{12}[f_1, f_2] = \mathcal{E}_{11,\tau_1} - f_1 + \mathcal{E}_{12,\tau_1} - f_1, \tag{2.122}$$

$$\mathcal{Q}_2[\{f_1, f_2\}] = \mathcal{Q}_{22}[f_2, f_2] + \mathcal{Q}_{21}[f_2, f_1] = \mathcal{E}_{22,\tau_2} - f_2 + \mathcal{E}_{21,\tau_2} - f_2 \tag{2.123}$$

where $\tau_i \in \{-1, 0, +1\}$.

The local equilibrium $\mathcal{E}_{ij,\tau_i}$ takes distinct forms depending on the type of the species, specified by τ_i. We distinguish between

$$\tau_i = \begin{cases} 0 & \text{for species } i \text{ being classic particles,} \\ +1 & \text{for species } i \text{ being fermions,} \\ -1 & \text{for species } i \text{ being bosons.} \end{cases}$$

A general formulation of the local equilibria is given in the following definition.

Definition 2.3.12 (Local equilibria for the quantum multi-species BGK model). *The target functions for the quantum multi-species BGK equations* (2.121) *are given by*

$$\mathcal{E}_{ii,\tau_i}[f_i] = \frac{1}{e^{m_i a_i \left|\frac{\mathbf{p}}{m_i} - \mathbf{b}_i\right|^2 + c_i} + \tau_i}, \tag{2.124a}$$

$$\mathcal{E}_{ij,\tau_i}[f_i, f_j] = \frac{1}{e^{m_i a \left|\frac{\mathbf{p}}{m_i} - \mathbf{b}\right|^2 + c_{ij}} + \tau_i} \tag{2.124b}$$

for $i, j = 1, 2,\ i \neq j$. The parameters $a_i, \mathbf{b}_i, c_i$ and $a, \mathbf{b}, c_{ij}$ depend on the corresponding distribution functions f_1 and f_2.

Using the above description, we obtain a universal formulation for a model describing the interactions of each of the following combinations:

1. fermion-fermion interactions: $\mathcal{E}_{ij,1} = \mathcal{F}_{ij}$. $(i, j = 1, 2)$
2. boson-boson interactions: $\mathcal{E}_{ij,-1} = \mathcal{B}_{ij}$. $(i, j = 1, 2)$
3. classical-classical interactions: $\mathcal{E}_{ij,0} = \mathcal{M}_{ij}$. $(i, j = 1, 2)$
4. fermion (f_1)-classical (f_2) interactions: $\mathcal{E}_{1j,1} = \mathcal{F}_{1j}$, $\mathcal{E}_{2j,0} = \mathcal{M}_{2j}$. $(j = 1, 2)$
5. classical (f_1)-boson (f_2) interactions: $\mathcal{E}_{1j,0} = \mathcal{M}_{1j}$, $\mathcal{E}_{2j,-1} = \mathcal{B}_{2j}$. $(j = 1, 2)$
6. fermion (f_1)-boson (f_2) interactions: $\mathcal{E}_{1j,1} = \mathcal{F}_{1j}$, $\mathcal{E}_{2j,-1} = \mathcal{B}_{2j}$. $(j = 1, 2)$

The target functions $\mathcal{E}_{ij,\tau_i}$ depend on the distribution functions f_i. Under appropriate assumptions on $N_i, \mathbf{P}_i, E_i$, they are uniquely defined which will be derived in the following sections.

Intra-species target functions

We seek a distribution function which conserves mass, momentum and energy for intra-species interactions. These conservation properties read in the quantum description

$$\int_{\mathbb{R}^3} \mathcal{E}_{ii,\tau_i}\, \mathrm{d}\mathbf{p} = N_i, \quad \int_{\mathbb{R}^3} \mathcal{E}_{ii,\tau_i} \mathbf{p}\, \mathrm{d}\mathbf{p} = \mathbf{P}_i, \quad \int_{\mathbb{R}^3} \mathcal{E}_{ii,\tau_i} \frac{|\mathbf{p}|^2}{2m_i}\, \mathrm{d}\mathbf{p} = E_i, \quad (i = 1, 2). \tag{2.125}$$

In order to simplify the notation, we introduce the function

$$j_{\tau_i}(x) = \frac{\int \frac{1}{e^{|\mathbf{p}|^2+x}+\tau_i}\,\mathrm{d}\mathbf{p}}{\left(\int \frac{|p|^2}{e^{|\mathbf{p}|^2+x}+\tau_i}\,\mathrm{d}\mathbf{p}\right)^{3/5}}, \tag{2.126}$$

and we define

$$l : \{+1, -1\} \to [-\infty, \infty], \quad l(x) = \begin{cases} l(+1) = -\infty, \\ l(-1) = 0. \end{cases}$$

The limit $j_{+1}(-\infty)$ is understood in the following sense:

$$j_{+1}(-\infty) = \lim_{x\to-\infty} j_{+1}(x).$$

Now, we can show that the intra-species target functions are well-posed, and moreover, give the corresponding parameters $a_i, \mathbf{b}_i, c_i$ at least implicitly.

Theorem 2.3.13 (Well-posedness of intra-species target functions for the quantum multi-species BGK model)**.** *For $\tau_i = 0$, we do not have any further requirements. For $\tau_i = \pm 1$, we assume*

$$\frac{N_i}{(2m_i E_i - \mathbf{P}_i^2/N_i)^{\frac{3}{5}}} \leq j_{\tau_i}(l(\tau_i)).$$

Then there exist unique parameters $a_i, \mathbf{b}_i, c_i$ for the target functions $\mathcal{E}_{ii,\tau_i}$ in (2.124a) *such that the model* (2.121) *satisfies the conservation constraints* (2.125) *during intra-species interactions.*

Proof. The statement for $\tau_i = 0$ is well-known and follows e.g. from Theorem 2.2.4 for $\nu_{ii} = 1$. The proof for $\tau_i = \pm 1$ can be found in [BY20]. The main challenge is to show that the equation $(i = 1, 2)$

$$j_{\tau_i}(c_i) = \frac{N_i}{(2m_i E_i - |\mathbf{P}_i|^2/N_i)^{\frac{3}{5}}},$$

has a unique solution c_i. With this implicitly defined value, we obtain a_i by

$$a_i = m_i \left(\int_{\mathbb{R}^3} \frac{1}{e^{|\mathbf{p}|^2+c_i} + \tau_i}\,\mathrm{d}\mathbf{p} \right)^{\frac{2}{3}} N_i^{-\frac{2}{3}},$$

and a direct computation gives

$$\mathbf{b}_i = \frac{\mathbf{P}_i}{m_i N_i}.$$

□

The ideas for the intra-species interactions can be extended to the inter-species case.

Inter-species target functions

For the inter-species interactions, we require the conservation properties

$$\begin{aligned}
&\int_{\mathbb{R}^3} \mathcal{E}_{12,\tau_1}\,\mathrm{d}\mathbf{p} = N_1, \qquad \int_{\mathbb{R}^3} \mathcal{E}_{21,\tau_2}\,\mathrm{d}\mathbf{p} = N_2,\\
&\int_{\mathbb{R}^3} \mathcal{E}_{12,\tau_1}\mathbf{p}\,\mathrm{d}\mathbf{p} + \int_{\mathbb{R}^3} \mathcal{E}_{21,\tau_2}\mathbf{p}\,\mathrm{d}\mathbf{p} = \mathbf{P}_1 + \mathbf{P}_2,\\
&\int_{\mathbb{R}^3} \mathcal{E}_{12,\tau_1}\frac{|\mathbf{p}|^2}{2m_1}\,\mathrm{d}\mathbf{p} + \int_{\mathbb{R}^3} \mathcal{E}_{21,\tau_2}\frac{|\mathbf{p}|^2}{2m_2}\,\mathrm{d}\mathbf{p} = E_1 + E_2.
\end{aligned} \tag{2.127}$$

As in the intra-species case, we make use of auxiliary functions

$$\tilde{h}_{\tau_i}(x) = \int_{\mathbb{R}^3} \frac{1}{e^{|\mathbf{p}|^2+x} + \tau_i}\,\mathrm{d}\mathbf{p}, \tag{2.128}$$

and

$$k_{\tau_i,\tau_j}(x,y) = \frac{m_1^{\frac{3}{2}} \int_{\mathbb{R}^3} \frac{1}{e^{|\mathbf{p}|^2+x}+\tau_i}\,\mathrm{d}\mathbf{p}}{\left(m_1^{\frac{3}{2}} \int_{\mathbb{R}^3} \frac{|\mathbf{p}|^2}{e^{|\mathbf{p}|^2+x}+\tau_i}\,\mathrm{d}\mathbf{p} + m_2^{\frac{3}{2}} \int_{\mathbb{R}^3} \frac{|\mathbf{p}|^2}{e^{|\mathbf{p}|^2+y}+\tau_j}\,\mathrm{d}\mathbf{p}\right)^{\frac{3}{5}}}. \tag{2.129}$$

Additionally, we define g_{τ_i,τ_j} as a composite function of k_{τ_i,τ_j} and $\tilde{h}_{\tau_i}^{-1}$ by

$$g_{\tau_i,\tau_j}(x) = k_{\tau_i,\tau_j}(x, y(x)) = \frac{m_1^{\frac{3}{2}} \int_{\mathbb{R}^3} \frac{1}{e^{|\mathbf{p}|^2+x}+\tau_i}\,\mathrm{d}\mathbf{p}}{\left(m_1^{\frac{3}{2}} \int_{\mathbb{R}^3} \frac{|\mathbf{p}|^2}{e^{|\mathbf{p}|^2+x}+\tau_i}\,\mathrm{d}\mathbf{p} + m_2^{\frac{3}{2}} \int_{\mathbb{R}^3} \frac{|\mathbf{p}|^2}{e^{|\mathbf{p}|^2+y(x)}+\tau_j}\,\mathrm{d}\mathbf{p}\right)^{\frac{3}{5}}}, \tag{2.130}$$

where $y(x)$ denotes

$$y(x) = \tilde{h}_{\tau_j}^{-1}\left(\frac{m_1^{\frac{3}{2}} N_2}{m_2^{\frac{3}{2}} N_1}\tilde{h}_{\tau_i}(x)\right).$$

Be aware that $\tilde{h}_{\tau_j}^{-1}$ always exists because $\tilde{h}_{\tau_i}$ is strictly decreasing.

Theorem 2.3.14 (Well-posedness of inter-species target functions for the quantum multi-species BGK model). *For $\tau_i = 0$, we do not have any further requirements. For $\tau_i = \pm 1$, we assume*

$$\frac{N_1}{\left(2E_1 + 2E_2 - \frac{|\mathbf{P}_1+\mathbf{P}_2|^2}{m_1 N_1 + m_2 N_2}\right)^{\frac{3}{5}}} \le g_{\tau_i,\tau_j}\left(\max\left\{l(\tau_i), \tilde{h}_{\tau_i}^{-1}\left(\frac{m_2^{\frac{3}{2}} N_1}{m_1^{\frac{3}{2}} N_2}\tilde{h}_{\tau_j}(l(\tau_j))\right)\right\}\right).$$

Then there exist unique parameters $a, \mathbf{b}, c_{12}, c_{21}$ for the target functions $\mathcal{E}_{12,\tau_1}$ and $\mathcal{E}_{21,\tau_2}$ in (2.124b) such that the model (2.121) satisfies the conservation constraints (2.127) during inter-species interactions.

Proof. The proof can be found in [BKPY21, BPW22]. We only give the relations which uniquely determine the equilibrium coefficients. An explicit calculation gives

$$\mathbf{P}_1 + \mathbf{P}_2 = \mathbf{b}(m_1 N_1 + m_2 N_2),$$

hence

$$\mathbf{b} = \frac{\mathbf{P}_1 + \mathbf{P}_2}{m_1 N_1 + m_2 N_2}. \tag{2.131}$$

Using

$$N_i = m_i^{\frac{3}{2}} a^{-\frac{3}{2}} \int_{\mathbb{R}^3} \frac{1}{e^{|\mathbf{p}|^2 + c_{ij}} + \tau_i} \, \mathrm{d}\mathbf{p} \tag{2.132}$$

and

$$\begin{aligned} &E_1 + E_2 \\ &= \frac{m_1^{\frac{3}{2}}}{2} a^{-\frac{5}{2}} \int_{\mathbb{R}^3} \frac{|\mathbf{p}|^2}{e^{m_1 a |\frac{\mathbf{p}}{m_1} - \mathbf{b}|^2 + c_{12}} + \tau_1} \, \mathrm{d}\mathbf{p} + \frac{m_2^{\frac{3}{2}}}{2} a^{-\frac{5}{2}} \int_{\mathbb{R}^3} \frac{|\mathbf{p}|^2}{e^{m_2 a |\frac{\mathbf{p}}{m_2} - \mathbf{b}|^2 + c_{21}} + \tau_2} \, \mathrm{d}\mathbf{p} \\ &\quad + \frac{1}{2}(m_1 N_1 + m_2 N_2)\mathbf{b}^2 \end{aligned}$$

yields

$$a = \left(\frac{m_1^{\frac{3}{2}} \int_{\mathbb{R}^3} \frac{|\mathbf{p}|^2}{e^{|\mathbf{p}|^2 + c_{12}} + \tau_1} \, \mathrm{d}\mathbf{p} + m_2^{\frac{3}{2}} \int_{\mathbb{R}^3} \frac{|\mathbf{p}|^2}{e^{|\mathbf{p}|^2 + c_{21}} + \tau_2} \, \mathrm{d}\mathbf{p}}{2E_1 + 2E_2 - \frac{|\mathbf{P}_1 + \mathbf{P}_2|^2}{m_1 N_1 + m_2 N_2}} \right)^{\frac{2}{5}}. \tag{2.133}$$

For a closed formula of a in (2.133), we need to define c_{12} and c_{21}. With the above equations and (2.129), we obtain

$$\frac{N_1}{\left(2E_1 + 2E_2 - \frac{|\mathbf{P}_1 + \mathbf{P}_2|^2}{m_1 N_1 + m_2 N_2}\right)^{\frac{3}{5}}} = k_{\tau_i, \tau_j}(c_{12}, c_{21}). \tag{2.134}$$

On top of that, it follows from (2.132) and (2.128) that

$$\frac{N_1}{N_2} = \frac{m_1^{\frac{3}{2}} \tilde{h}_{\tau_1}(c_{12})}{m_2^{\frac{3}{2}} \tilde{h}_{\tau_2}(c_{21})}.$$

It can be shown that g_{τ_i, τ_j} is strictly monotonically decreasing [BKPY21]. It follows that c_{12}, c_{21} can be defined as unique solutions of the above relations, and

$$c_{21} = \tilde{h}_{\tau_2}^{-1} \left(\frac{m_1^{\frac{3}{2}} N_2}{m_2^{\frac{3}{2}} N_1} \tilde{h}_{\tau_1}(c_{12}) \right).$$

□

We study the properties of the model in the following.

Consistency of the model

In this section, we present properties of the quantum multi-species BGK model (2.121). First of all, the model satisfies the conservation properties and an $\mathcal{H}$-Theorem. This is stated in the following theorems and proven in [BKPY21].

Theorem 2.3.15 (Conservation properties for the quantum multi-species BGK model). *Let the equilibrium coefficients be chosen as given in the proofs of Theorems 2.3.13 and 2.3.14. Then the model* (2.121) *conserves mass, total momentum and total energy if the assumptions in Theorems 2.3.13 and 2.3.14 are fulfilled.*

Proof. This is an immediate consequence of Theorems 2.3.13 and 2.3.14. □

Theorem 2.3.16 ($\mathcal{H}$-Theorem for the quantum multi-species BGK model). *Let us denote the function $h_\tau(z) = z\log(z) - z + \tau(1-\tau z)\log(1-\tau z) - \tau(1-\tau z)$ and the total entropy $\mathcal{H}[\{f_1, f_2\}] = \int (h_{\tau_1}(f_1) + h_{\tau_2}(f_2))\, \mathrm{d}\mathbf{p}$. With the choice of equilibrium coefficients given in the proofs of Theorems 2.3.13 and 2.3.14, the quantum BGK model for gas mixtures* (2.121) *satisfies the following entropy inequality*

$$\partial_t \mathcal{H}[\{f_1, f_2\}] + \nabla_{\mathbf{x}} \cdot \left[\int \mathbf{v}(h_{\tau_1}(f_1) + h_{\tau_2}(f_2))\, \mathrm{d}\mathbf{p} \right] \leq 0. \tag{2.135}$$

The equality in (2.135) *is characterized by f_i being Fermi-Dirac distribution functions for fermions, Bose-Einstein distributions functions for bosons and Maxwellian distribution functions for classical particles. In all cases, the equilibrium distributions share the same a and $\mathbf{b}$.*

Proof. We recall the proof from [BKPY21]. A direct calculation gives

$$\begin{aligned}\partial_t \mathcal{H}[\{f_1, f_2\}] + \nabla_{\mathbf{x}} \cdot \left[\int \mathbf{v}(h_{\tau_1}(f_1) + h_{\tau_2}(f_2))\, \mathrm{d}\mathbf{p} \right] &= \sum_{i=1}^{2} \int \mathcal{Q}_i[\{f_i\}] h'_{\tau_i}(f_i)\, \mathrm{d}\mathbf{v} \\ &= D_{11} + D_{12} + D_{21} + D_{22}\end{aligned}$$

with the dissipation terms

$$D_{ij} = \int (\mathcal{E}_{ij,\tau_i} - f_i) h'_{\tau_i}(f_i)\, \mathrm{d}\mathbf{p}$$

and the derivative

$$h'_\tau(z) = \log \frac{z}{1-\tau z}.$$

We show that $D_{ii} \leq 0$ and $D_{12} + D_{21} \leq 0$ which proves (2.135).

Due to the conservation during intra-species interactions (2.125), it is

$$\int (\mathcal{E}_{ii,\tau_i} - f_i) h'_{\tau_i}(\mathcal{E}_{ii,\tau_i})\, \mathrm{d}\mathbf{p} = 0.$$

It follows

$$\begin{aligned} D_{ii} &= \int (\mathcal{E}_{ii,\tau_i} - f_i) h'_{\tau_i}(f_i) \,\mathrm{d}\mathbf{v} - \int (\mathcal{E}_{ii,\tau_i} - f_i) h'_{\tau_i}(\mathcal{E}_{ii,\tau_i}) \,\mathrm{d}\mathbf{p} \\ &= \int (\mathcal{E}_{ii,\tau_i} - f_i)(h'_{\tau_i}(f_i) - h'_{\tau_i}(\mathcal{E}_{ii,\tau_i})) \,\mathrm{d}\mathbf{p} \leq 0. \end{aligned} \tag{2.136}$$

The last inequality holds because $h_{-1}(z)$, $h_0(z)$ are increasing functions for $z \in [0, \infty)$, and $h_1(z)$ increases for $0 < z < 1$, resulting in

$$(z - y)(h_\tau(y) - h_\tau(z)) \leq 0 \tag{2.137}$$

for $y, z \in \mathbb{R}_0^+$ $(\tau = 0, -1)$ and for $0 < y, z < 1$ $(\tau = +1)$, respectively. Additionally, (2.137) is an equality if and only if $y = z$. Applied to (2.136), $D_{ii} = 0$ if and only if $f_i = \mathcal{E}_{ii,\tau_i}$.

For the mixed dissipation terms, we use the conservation during inter-species interactions (2.127) and observe

$$I := \int (\mathcal{E}_{12,\tau_1} - f_1) h'_{\tau_1}(\mathcal{E}_{12,\tau_1}) \,\mathrm{d}\mathbf{p} + \int (\mathcal{E}_{21,\tau_2} - f_2) h'_{\tau_2}(\mathcal{E}_{21,\tau_2}) \,\mathrm{d}\mathbf{p} = 0.$$

Adding this vanishing term to the dissipation terms yields

$$\begin{aligned} D_{12} + D_{21} &= D_{12} + D_{21} - I \\ &= \int (\mathcal{E}_{12,\tau_1} - f_1)(h'_{\tau_1}(f_1) - h'_{\tau_1}(\mathcal{E}_{12,\tau_1})) \,\mathrm{d}\mathbf{p} \\ &\quad + \int (\mathcal{E}_{21,\tau_2} - f_2)(h'_{\tau_2}(f_2) - h'_{\tau_2}(\mathcal{E}_{21,\tau_2})) \,\mathrm{d}\mathbf{p} \\ &\overset{(2.137)}{\leq} 0 \end{aligned}$$

with equality if and only if $f_1 = \mathcal{E}_{12,\tau_1}$ and $f_2 = \mathcal{E}_{21,\tau_2}$ using the same arguments as above. It follows that the equilibrium distributions share the same a and $\mathbf{b}$. □

Moreover, one can prove that the distribution function in the fermion case remains bounded by 1 for all times $t \geq 0$.

Lemma 2.3.17. *Let f_i be a distribution function for fermions and $f_i(\mathbf{x}, \mathbf{p}, 0) < 1$. Then we have $f_i(\mathbf{x}, \mathbf{p}, t) < 1$ for $t \geq 0$.*

Proof. We paraphrase the proof presented in [BKPY21]. We integrate the evolution equation (2.121) along the characteristic and obtain

$$f_i(\mathbf{x}, \mathbf{p}, t) = e^{-2t} f_i\left(\mathbf{x} - \frac{\mathbf{p}}{m_i} t, \mathbf{p}, 0\right) + \int_0^t e^{2(s-t)} [\mathcal{F}_{ii} + \mathcal{F}_{ij}](\mathbf{x} + (s - t)\mathbf{p}, \mathbf{p}, s) \,\mathrm{d}s.$$

By definition, it is $0 < \mathcal{F}_{ii}, \mathcal{F}_{ij} < 1$ for all $(\mathbf{x}, \mathbf{p}, t)$, and hence

$$\begin{aligned} f_i(\mathbf{x}, \mathbf{p}, t) &\leq e^{-2t} f_i\left(\mathbf{x} - \frac{\mathbf{p}}{m_i}t, \mathbf{p}, 0\right) + \int_0^t 2e^{2(s-t)}\, \mathrm{d}s \\ &= e^{-2t} f_i\left(\mathbf{x} - \frac{\mathbf{p}}{m_i}t, \mathbf{p}, 0\right) + (1 - e^{-2t}) \\ &< e^{-2t} + 1 - e^{-2t} = 1. \end{aligned}$$

□

We are now interested in the hydrodynamic behavior of the presented equations.

Macroscopic equations

From the quantum multi-species BGK model (2.121), one can derive the following macroscopic equations.

Theorem 2.3.18 (Macroscopic equations for the quantum multi-species BGK model). *Let $\{f_1, f_2\}$ be a solution to (2.121), then we obtain the following formal conservation laws*

$$\begin{aligned} \partial_t N_1 + \nabla_{\mathbf{x}} \cdot \frac{\mathbf{P}_1}{m_1} &= 0, \\ \partial_t N_2 + \nabla_{\mathbf{x}} \cdot \frac{\mathbf{P}_2}{m_2} &= 0, \\ \partial_t \mathbf{P}_1 + \nabla_{\mathbf{x}} \cdot \int \mathbf{p} \otimes \frac{\mathbf{p}}{m_1} f_1(\mathbf{p})\, \mathrm{d}\mathbf{p} &= \mathbf{P}_{12} - \mathbf{P}_1, \\ \partial_t \mathbf{P}_2 + \nabla_{\mathbf{x}} \cdot \int \mathbf{p} \otimes \frac{\mathbf{p}}{m_2} f_2(\mathbf{p})\, \mathrm{d}\mathbf{p} &= \mathbf{P}_{21} - \mathbf{P}_2, \\ \partial_t E_1 + \nabla_{\mathbf{x}} \cdot \int \frac{|\mathbf{p}|^2}{2m_1} \frac{\mathbf{p}}{m_1}\, \mathrm{d}\mathbf{p} &= E_{12} - E_1, \\ \partial_t E_2 + \nabla_{\mathbf{x}} \cdot \int \frac{|\mathbf{p}|^2}{2m_2} \frac{\mathbf{p}}{m_2}\, \mathrm{d}\mathbf{p} &= E_{21} - E_2, \end{aligned} \tag{2.138}$$

where the exchange terms of momentum are given by

$$\mathbf{P}_{12} - \mathbf{P}_1 = -(\mathbf{P}_{21} - \mathbf{P}_2) = \frac{m_1 N_1 m_2 N_2}{m_1 N_1 + m_2 N_2}\left(\frac{\mathbf{P}_2}{m_2 N_2} - \frac{\mathbf{P}_1}{m_1 N_1}\right).$$

We define the function

$$H_{\tau_i}(c) = \int \frac{|\mathbf{p}|^2}{e^{|\mathbf{p}|^2 + c} + \tau_i}\, \mathrm{d}\mathbf{p}$$

and obtain for the exchange of energy

$$\begin{aligned} E_{12} - E_1 &= \frac{1}{2}\frac{m_1 N_1 |\mathbf{P}_1 + \mathbf{P}_2|^2}{(m_1 N_1 + m_2 N_2)^2} + \frac{(E_1 + E_2) - \frac{1}{2}\frac{|\mathbf{P}_1 + \mathbf{P}_2|^2}{m_1 N_1 + m_2 N_2}}{m_1^{3/2} H_{\tau_1}(c_{12}) + m_2^{3/2} H_{\tau_2}(c_{21})} m_1^{3/2} H_{\tau_1}(c_{12}) - E_1 \\ &= -(E_{21} - E_2). \end{aligned} \tag{2.139}$$

Proof. We multiply the first equation of (2.121) with $(1, \mathbf{p}, \frac{|\mathbf{p}|^2}{2m_1})$, and the second one with $(1, \mathbf{p}, \frac{|\mathbf{p}|^2}{2m_2})$. Integration with respect to the momentum $\mathbf{p}$ yields the LHS of (2.138) in a straight-forward way.

The exchange of momentum can be computed by multiplying (2.121) with $\mathbf{p}$ and integrating with respect to $\mathbf{p}$. That is

$$\begin{aligned}\mathbf{P}_{12} - \mathbf{P}_1 &= \int_{\mathbb{R}^3} \frac{\mathbf{p}}{e^{m_1 a\left|\frac{\mathbf{p}}{m_1} - \mathbf{b}\right|^2 + c_{12}} + \tau_1} \,\mathrm{d}\mathbf{p} - \mathbf{P}_1 = \int_{\mathbb{R}^3} \frac{\mathbf{p} + m_1\mathbf{b}}{e^{a|\mathbf{p}|^2 + c_{12}} + \tau_1} \,\mathrm{d}\mathbf{p} - \mathbf{P}_1 \\ &= m_1\mathbf{b}N_1 - \mathbf{P}_1 \overset{(2.131)}{=} m_1 N_1 \frac{\mathbf{P}_1 + \mathbf{P}_2}{m_1N_1 + m_2N_2} - \mathbf{P}_1 \\ &= \frac{m_1N_1m_2N_2}{m_1N_1 + m_2N_2}\left(\frac{\mathbf{P}_2}{m_2N_2} - \frac{\mathbf{P}_1}{m_1N_1}\right) = -(\mathbf{P}_{21} - \mathbf{P}_2).\end{aligned} \tag{2.140}$$

For the exchange of energy, we start with a change of variables (similar as in [BY20, Section 2]) leading to

$$\begin{aligned}E_{12} - \frac{1}{2}\frac{|\mathbf{P}_{12}|^2}{m_1N_1} &= \frac{1}{m_1}\int_{\mathbb{R}^3} \frac{|\mathbf{p}|^2}{e^{m_1 a\left|\frac{\mathbf{p}}{m_1} - \mathbf{b}\right|^2 + c_{12}} + \tau_1} \,\mathrm{d}\mathbf{p} - \frac{1}{2}\frac{|\mathbf{P}_{12}|^2}{m_1N_1} \\ &= \frac{1}{2}a^{-5/2}m_1^{3/2}H_{\tau_1}(c_{12}).\end{aligned} \tag{2.141}$$

Inserting the expression for a from Theorem 2.3.14 in (2.133) yields

$$E_{12} - \frac{1}{2}\frac{|\mathbf{P}_{12}|^2}{m_1N_1} = \frac{(E_1 + E_2) - \frac{1}{2}\frac{|\mathbf{P}_1 + \mathbf{P}_2|^2}{m_1N_1 + m_2N_2}}{m_1^{3/2}H_{\tau_1}(c_{12}) + m_2^{3/2}H_{\tau_2}(c_{21})} m_1^{3/2}H_{\tau_1}(c_{12}).$$

Replacing $\mathbf{P}_{12}$ with Theorem 2.3.14 concludes to

$$\begin{aligned}E_{12} - E_1 &= \frac{1}{2}\frac{m_1N_1|\mathbf{P}_1 + \mathbf{P}_2|^2}{(m_1N_1 + m_2N_2)^2} + \frac{(E_1 + E_2) - \frac{1}{2}\frac{|\mathbf{P}_1 + \mathbf{P}_2|^2}{m_1N_1 + m_2N_2}}{m_1^{3/2}H_{\tau_1}(c_{12}) + m_2^{3/2}H_{\tau_2}(c_{21})} m_1^{3/2}H_{\tau_1}(c_{12}) - E_1 \\ &= -(E_{21} - E_2).\end{aligned}$$

The proof was done by Pirner in [BPW22]. □

In general, the exchange of energy in the macroscopic equations (2.138) cannot be expressed explicitly because H_{τ_i} might not be computable analytically. However, considering only classical interactions, the exchange terms can be expressed by closed formulae.

Remark 2.3.19. *In the classical-classical case $\tau_1 = \tau_2 = 0$, the relationship* (2.134) *in Theorem 2.3.14 simplifies to*

$$\frac{N_1}{N_2} = \frac{m_1^{\frac{3}{2}}\tilde{h}_0(c_{12})}{m_2^{\frac{3}{2}}\tilde{h}_0(c_{21})} = \frac{m_1^{\frac{3}{2}}\int_{\mathbb{R}^3}\frac{1}{e^{|\mathbf{p}|^2 + c_{12}} + 0}\,\mathrm{d}\mathbf{p}}{m_2^{\frac{3}{2}}\int_{\mathbb{R}^3}\frac{1}{e^{|\mathbf{p}|^2 + c_{21}} + 0}\,\mathrm{d}\mathbf{p}} = \frac{m_1^{3/2}}{m_2^{3/2}}\frac{e^{-c_{12}}}{e^{-c_{21}}}.$$

Further, computing the integrals

$$\frac{m_1^{3/2} H_0(c_{12})}{m_1^{3/2} H_0(c_{12}) + m_2^{3/2} H_0(c_{21})} = \frac{m_1^{3/2} e^{-c_{12}}}{m_1^{3/2} e^{-c_{12}} + m_2^{3/2} e^{-c_{21}}}$$
$$= \frac{m_1^{3/2}}{m_1^{3/2} e^{-c_{12}} + m_1^{3/2} \frac{N_2}{N_1} e^{-c_{12}}} = \frac{N_1}{N_1+N_2},$$

we obtain

$$\begin{aligned} E_{12} - E_1 &= \frac{N_1 N_2}{N_1+N_2} \left(\frac{E_2}{N_2} - \frac{E_1}{N_1} + \frac{m_1 - m_2}{(m_1 N_1 + m_2 N_2)^2} \frac{1}{2} |\mathbf{P}_1 + \mathbf{P}_2|^2 \right) \\ &= \frac{N_1 N_2}{N_1+N_2} \left(\frac{E_2}{N_2} - \frac{1}{2} \frac{|\mathbf{P}_2|^2}{m_2 N_2^2} - \frac{E_1}{N_1} + \frac{|\mathbf{P}_1|^2}{m_1 N_1^2} \right. \\ &\quad + m_1 m_2 \frac{m_1 N_1^2 + m_2 N_2^2}{(m_1 N_1 + m_2 N_2)^2} \frac{1}{2} \left(\frac{|\mathbf{P}_2|^2}{m_2^2 N_2^2} - \frac{|\mathbf{P}_1|^2}{m_2^2 N_1^2} \right) \\ &\quad \left. + \frac{m_1 m_2 N_1 N_2}{(m_1 N_1 + m_2 N_2)^2} \left(\frac{\mathbf{P}_2}{m_2 N_2} - \frac{\mathbf{P}_1}{m_1 N_1} \right) \cdot \left(\frac{\mathbf{P}_1}{N_1} + \frac{\mathbf{P}_2}{N_2} \right) \right). \end{aligned}$$

We can also specify decay rates for macroscopic quantities.

Decay rates for the mean velocities and temperatures in the space homogeneous case

The $\mathcal{H}$-Theorem states that both species share the same mean velocity in equilibrium. More precisely, we can compute an explicit rate with which the mean velocities converge.

Theorem 2.3.20 (Estimates for the mean velocities for the quantum multi-species BGK model)**.** *In the space homogeneous case of* (2.121), *we have the following convergence rate for the momenta:*

$$\frac{\mathbf{P}_1}{m_1 N_1} - \frac{\mathbf{P}_2}{m_2 N_2} = e^{-t} \left(\frac{\mathbf{P}_1(0)}{m_1 N_1} - \frac{\mathbf{P}_2(0)}{m_2 N_2} \right). \tag{2.142}$$

Proof. Since N_i is constant in the space homogeneous case (2.138), we have

$$\begin{aligned} \partial_t \left(\frac{\mathbf{P}_1}{m_1 N_1} \right) &= \frac{1}{m_1 N_1} \partial_t \mathbf{P}_1 = \frac{1}{m_1 N_1} (\mathbf{P}_{12} - \mathbf{P}_1) \\ &\overset{(2.140)}{=} \frac{m_2 N_2}{m_1 N_1 + m_2 N_2} \left(\frac{\mathbf{P}_2}{m_2 N_2} - \frac{\mathbf{P}_1}{m_1 N_1} \right). \end{aligned} \tag{2.143}$$

In a similar way, we compute

$$\partial_t \left(\frac{\mathbf{P}_2}{m_2 N_2} \right) = \frac{m_1 N_1}{m_1 N_1 + m_2 N_2} \left(\frac{\mathbf{P}_1}{m_1 N_1} - \frac{\mathbf{P}_2}{m_2 N_2} \right). \tag{2.144}$$

If we subtract the two equations (2.143)–(2.144), we obtain

$$\partial_t \left(\frac{\mathbf{P}_1}{m_1 N_1} - \frac{\mathbf{P}_2}{m_2 N_2} \right) = - \left(\frac{\mathbf{P}_1}{m_1 N_1} - \frac{\mathbf{P}_2}{m_2 N_2} \right)$$

and conclude

$$\frac{\mathbf{P}_1}{m_1 N_1} - \frac{\mathbf{P}_2}{m_2 N_2} = e^{-t} \left(\frac{\mathbf{P}_1(0)}{m_1 N_1} - \frac{\mathbf{P}_2(0)}{m_2 N_2} \right).$$

The proof was done by Pirner in [BPW22]. □

Remark 2.3.21. *Equivalently to* (2.142), *one can also write*

$$\mathbf{b}_1 - \mathbf{b}_2 = e^{-t}(\mathbf{b}_1(0) - \mathbf{b}_2(0))$$

using the relationship $\mathbf{P}_i = m_i N_i \mathbf{b}_i$.

We continue with the convergence rates of the kinetic temperatures $\frac{3}{2}T_i = \frac{E_i}{N_i} - \frac{1}{2}\frac{|\mathbf{P}_i|^2}{m_i N_i^2}$.

Theorem 2.3.22 (Estimates for the kinetic temperatures for the quantum multi-species BGK model). *In the space homogeneous case of* (2.121), *it is*

$$\begin{aligned}
&\left(\frac{E_1}{N_1} - \frac{1}{2}\frac{|\mathbf{P}_1|^2}{m_1 N_1^2} \right) - \left(\frac{E_2}{N_2} - \frac{1}{2}\frac{|\mathbf{P}_2|^2}{m_2 {N_2}^2} \right) \\
&\quad = e^{-t} \left(\left(\frac{E_1(0)}{N_1} - \frac{1}{2}\frac{|\mathbf{P}_1(0)|^2}{m_1 N_1^2} \right) - \left(\frac{E_2(0)}{N_2} - \frac{1}{2}\frac{|\mathbf{P}_2(0)|^2}{m_2 N_2^2} \right) \right) \\
&\qquad + \frac{1}{2} m_1 m_2 \frac{m_2 N_2^2 - m_1 N_1^2}{(m_1 N_1 + m_2 N_2)^2} e^{-t}(1 - e^{-t}) \left| \frac{\mathbf{P}_2(0)}{m_2 N_2} - \frac{\mathbf{P}_1(0)}{m_1 N_1} \right|^2 \\
&\qquad + \left(E_1(0) + E_2(0) - \frac{1}{2}\frac{|\mathbf{P}_1(0) + \mathbf{P}_2(0)|^2}{m_1 N_1 + m_2 N_2} \right) e^{-t} \\
&\qquad\quad \cdot \int_0^t e^s \left[\frac{\frac{m_1^{3/2}}{N_1} H_{\tau_1}(c_{12}(s)) - \frac{m_2^{3/2}}{N_2} H_{\tau_2}(c_{21}(s))}{m_1^{3/2} H_{\tau_1}(c_{12}(s)) + m_2^{3/2} H_{\tau_2}(c_{21}(s))} \right] ds.
\end{aligned} \tag{2.145}$$

Proof. Using (2.138) and inserting (2.143), we compute

$$\begin{aligned}
\partial_t \left(\frac{E_1}{N_1} - \frac{1}{2}\frac{|\mathbf{P}_1|^2}{m_1 N_1^2} \right) &= \partial_t \left(\frac{E_1}{N_1} \right) - \frac{\mathbf{P}_1}{N_1} \partial_t \left(\frac{\mathbf{P}_1}{m_1 N_1} \right) \\
&= \frac{E_{12}}{N_1} - \frac{E_1}{N_1} - \frac{\mathbf{P}_1}{N_1} \frac{m_2 N_2}{m_1 N_1 + m_2 N_2} \left(\frac{\mathbf{P}_2}{m_2 N_2} - \frac{\mathbf{P}_1}{m_1 N_1} \right).
\end{aligned} \tag{2.146}$$

We plug (2.139) in and obtain

$$\begin{aligned}
&\partial_t \left(\frac{E_1}{N_1} - \frac{1}{2} \frac{|\mathbf{P}_1|^2}{m_1 N_1^2} \right) \\
&= \frac{1}{2} \frac{m_1 |\mathbf{P}_1 + \mathbf{P}_2|^2}{(m_1 N_1 + m_2 N_2)^2} + \frac{(E_1 + E_2) - \frac{1}{2} \frac{|\mathbf{P}_1+\mathbf{P}_2|^2}{m_1 N_1 + m_2 N_2}}{m_1^{3/2} H_{\tau_1}(c_{12}) + m_2^{3/2} H_{\tau_2}(c_{21})} \frac{m_1^{3/2} H_{\tau_1}(c_{12})}{N_1} \\
&\qquad - \frac{E_1}{N_1} - \frac{\mathbf{P}_1}{N_1} \frac{m_2 N_2}{m_1 N_1 + m_2 N_2} \left(\frac{\mathbf{P}_2}{m_2 N_2} - \frac{\mathbf{P}_1}{m_1 N_1} \right).
\end{aligned} \tag{2.147}$$

An analogous expression can be derived for species 2. Subtracting both leads to

$$\begin{aligned}
&\partial_t \left(\left(\frac{E_1}{N_1} - \frac{1}{2} \frac{|\mathbf{P}_1|^2}{m_1 N_1^2} \right) - \left(\frac{E_2}{N_2} - \frac{|\mathbf{P}_2|^2}{m_2 N_2^2} \right) \right) \\
&= \frac{E_2}{N_2} - \frac{E_1}{N_1} + \varpi + \frac{(E_1 + E_2) - \frac{1}{2} \frac{|\mathbf{P}_1+\mathbf{P}_2|^2}{m_1 N_1 + m_2 N_2}}{m_1^{3/2} H_{\tau_1}(c_{12}) + m_2^{3/2} H_{\tau_2}(c_{21})} \left[\frac{m_1^{3/2} H_{\tau_1}(c_{12})}{N_1} - \frac{m_2^{3/2} H_{\tau_2}(c_{21})}{N_2} \right]
\end{aligned}$$

with

$$\begin{aligned}
\varpi =& \frac{\frac{1}{2}(m_1 - m_2) m_1 N_1^2 + m_2 N_2 (m_1 N_1 + m_2 N_2)}{(m_1 N_1 + m_2 N_2)^2} \frac{|\mathbf{P}_1|^2}{m_1 N_1^2} \\
&+ \frac{(m_1 N_1^2 - m_2 N_2^2)}{(m_1 N_1 + m_2 N_2)^2 N_1 N_2} \mathbf{P}_1 \cdot \mathbf{P}_2 \\
&+ \frac{\frac{1}{2}(m_1 - m_2) m_2 N_2^2 - m_1 N_1 (m_1 N_1 + m_2 N_2)}{(m_1 N_1 + m_2 N_2)^2} \frac{|\mathbf{P}_2|^2}{m_2 N_2^2} \\
=& \frac{1}{2} m_1 m_2 \frac{m_2 N_2^2 - m_1 N_1^2}{(m_1 N_1 + m_2 N_2)^2} \left| \frac{\mathbf{P}_1}{m_1 N_1} - \frac{\mathbf{P}_2}{m_2 N_2} \right|^2 + \frac{1}{2} \frac{|\mathbf{P}_1|^2}{m_1 N_1^2} - \frac{1}{2} \frac{|\mathbf{P}_2|^2}{m_2 N_2^2}.
\end{aligned}$$

This can be rewritten to

$$\partial_t \left(\left(\frac{E_1}{N_1} - \frac{1}{2} \frac{|\mathbf{P}_1|^2}{m_1 N_1^2} \right) - \left(\frac{E_2}{N_2} - \frac{1}{2} \frac{|\mathbf{P}_2|^2}{m_2 N_2^2} \right) \right) = T_E + T_P + T_H$$

where

$$\begin{aligned}
T_E(t) &= - \left(\left(\frac{E_1}{N_1} - \frac{|\mathbf{P}_1|^2}{m_1 N_1^2} \right) - \left(\frac{E_2}{N_2} - \frac{|\mathbf{P}_2|^2}{m_2 N_2^2} \right) \right), \\
T_P(t) &= \frac{1}{2} m_1 m_2 \frac{m_2 N_2^2 - m_1 N_1^2}{(m_1 N_1 + m_2 N_2)^2} \left| \frac{\mathbf{P}_1}{m_1 N_1} - \frac{\mathbf{P}_2}{m_2 N_2} \right|^2, \\
T_H(t) &= \frac{(E_1 + E_2) - \frac{1}{2} \frac{|\mathbf{P}_1+\mathbf{P}_2|^2}{m_1 N_1 + m_2 N_2}}{m_1^{3/2} H_{\tau_1}(c_{12}) + m_2^{3/2} H_{\tau_2}(c_{21})} \left[\frac{m_1^{3/2} H_{\tau_1}(c_{12})}{N_1} - \frac{m_2^{3/2} H_{\tau_2}(c_{21})}{N_2} \right].
\end{aligned}$$

Now, Duhamels formula gives

$$\left(\frac{E_1}{N_1}-\frac{1}{2}\frac{|\mathbf{P}_1|^2}{m_1 N_1^2}\right)-\left(\frac{E_2}{N_2}-\frac{1}{2}\frac{|\mathbf{P}_2|^2}{m_2 N_2^2}\right)=T_E^*+T_P^*+T_H^*$$

with

$$T_E^*=e^{-t}T_E(0)=e^{-t}\left(\left(\frac{E_1(0)}{N_1}-\frac{1}{2}\frac{|\mathbf{P}_1(0)|^2}{m_1 N_1^2}\right)-\left(\frac{E_2(0)}{N_2}-\frac{1}{2}\frac{|\mathbf{P}_2(0)|^2}{m_2 N_2^2}\right)\right)$$

and

$$T_H^*=\left(E_1(0)+E_2(0)-\frac{1}{2}\frac{|\mathbf{P}_1(0)+\mathbf{P}_2(0)|^2}{m_1N_1+m_2N_2}\right)e^{-t}$$
$$\cdot\int_0^t e^s\left[\frac{\frac{m_1^{3/2}}{N_1}H_{\tau_1}(c_{12}(s))-\frac{m_2^{3/2}}{N_2}H_{\tau_2}(c_{21}(s))}{m_1^{3/2}H_{\tau_1}(c_{12}(s))+m_2^{3/2}H_{\tau_2}(c_{21}(s))}\right]\mathrm{d}s.$$

Using Theorem 2.3.20 yields

$$\begin{aligned}T_P^*&=\frac{1}{2}m_1m_2\frac{m_2N_2^2-m_1N_1^2}{(m_1N_1+m_2N_2)^2}e^{-t}\int_0^t e^s\left|\frac{\mathbf{P}_2(s)}{m_2N_2}-\frac{\mathbf{P}_1(s)}{m_1N_1}\right|^2\mathrm{d}s\\&=\frac{1}{2}m_1m_2\frac{m_2N_2^2-m_1N_1^2}{(m_1N_1+m_2N_2)^2}e^{-t}\int_0^t e^s e^{-2s}\left|\frac{\mathbf{P}_2(0)}{m_2N_2}-\frac{\mathbf{P}_1(0)}{m_1N_1}\right|^2\mathrm{d}s\\&=\frac{1}{2}m_1m_2\frac{m_2N_2^2-m_1N_1^2}{(m_1N_1+m_2N_2)^2}e^{-t}(1-e^{-t})\left|\frac{\mathbf{P}_2(0)}{m_2N_2}-\frac{\mathbf{P}_1(0)}{m_1N_1}\right|^2,\end{aligned}$$

which finishes the proof. The proof was done by Pirner in [BPW22]. □

The complicated formula (2.145) can be simplified in the case of classical-classical interactions.

Remark 2.3.23. *We continue Remark 2.3.19. In the classical-classical case $\tau_1=\tau_2=0$, we get*

$$\frac{N_1}{N_2}=\frac{m_1^{3/2}}{m_2^{3/2}}\frac{e^{-c_{12}}}{e^{-c_{21}}},$$

and we are able to explicitly compute the bracket term of T_H^*:

$$\begin{aligned}&\left[\frac{\frac{m_1^{3/2}}{N_1}H_{\tau_1}(c_{12}(s))-\frac{m_2^{3/2}}{N_2}H_{\tau_2}(c_{21}(s))}{m_1^{3/2}H_{\tau(1)}(c_{12}(s))+m_2^{3/2}H_{\tau(2)}(c_{21}(s))}\right]=\frac{\frac{m_1^{3/2}}{N_1}\frac{3}{2}e^{-c_{12}}\pi^{3/2}-\frac{m_2^{3/2}}{N_2}\frac{3}{2}e^{-c_{21}}\pi^{3/2}}{m_1^{3/2}\frac{3}{2}e^{-c_{12}}\pi^{3/2}+m_2^{3/2}\frac{3}{2}e^{-c_{21}}\pi^{3/2}}\\&=\frac{\frac{m_1^{3/2}}{N_1}e^{-c_{12}}-\frac{m_2^{3/2}}{N_2}e^{-c_{21}}}{m_1^{3/2}e^{-c_{12}}+m_2^{3/2}e^{-c_{21}}}=\frac{\frac{m_1^{3/2}}{N_1}e^{-c_{12}}-\frac{m_2^{3/2}}{N_1}\frac{m_1^{3/2}}{m_2^{3/2}}e^{-c_{21}}}{m_1^{3/2}e^{-c_{12}}+m_1^{3/2}\frac{N_1}{N_2}e^{-c_{12}}}\\&=0.\end{aligned}$$

This means that in the classical-classical case, the decay rate for the kinetic temperatures (2.145) *simplifies to*

$$\left(\frac{E_1}{N_1} - \frac{1}{2}\frac{|\mathbf{P}_1|^2}{m_1 N_1^2}\right) - \left(\frac{E_2}{N_2} - \frac{1}{2}\frac{|\mathbf{P}_2|^2}{m_2 N_2^2}\right) = T_E^* + T_P^*$$

which equals the temperature rate for the KPP model in Theorem 2.1.44.

This section sets collision frequencies $\nu_{ij} = 1$. Actually, it is not clear how collision kernels in the quantum Boltzmann model look like [Vil02] which led to collision frequencies in the BGK model.

The following observation concludes this section and illustrates that the mean velocities and temperature converge the faster the larger the possible collision frequencies are.

Remark 2.3.24. *If we include constant collision frequencies ν_{ij} into the model with the condition*

$$\nu_{12} = \nu_{21},$$

i.e.

$$\begin{aligned}
\partial_t f_1 &= \nu_{11}(\mathcal{E}_{11} - f_1) + \nu_{12}(\mathcal{E}_{12} - f_1),\\
\partial_t f_2 &= \nu_{22}(\mathcal{E}_{22} - f_2) + \nu_{12}(\mathcal{E}_{21} - f_2),
\end{aligned}$$

then we can repeat all computations for the decay rates in Theorems 2.3.20 and 2.3.22 in the same way resulting in

$$\frac{\mathbf{P}_1}{m_1 N_1} - \frac{\mathbf{P}_2}{m_2 N_2} = e^{-\nu_{12} t}\left(\frac{\mathbf{P}_1(0)}{m_1 N_1} - \frac{\mathbf{P}_2(0)}{m_2 N_2}\right)$$

and

$$\begin{aligned}
&\left(\frac{E_1}{N_1} - \frac{1}{2}\frac{|\mathbf{P}_1|^2}{m_1 N_1^2}\right) - \left(\frac{E_2}{N_2} - \frac{1}{2}\frac{|\mathbf{P}_2|^2}{m_2 {N_2}^2}\right)\\
&= e^{-\nu_{12} t}\left(\left(\frac{E_1(0)}{N_1} - \frac{1}{2}\frac{|\mathbf{P}_1(0)|^2}{m_1 N_1^2}\right) - \left(\frac{E_2(0)}{N_2} - \frac{1}{2}\frac{|\mathbf{P}_2(0)|^2}{m_2 N_2^2}\right)\right)\\
&\quad + \frac{1}{2} m_1 m_2 \frac{m_2 N_2^2 - m_1 N_1^2}{(m_1 N_1 + m_2 N_2)^2} e^{-\nu_{12} t}(1 - e^{-\nu_{12} t}) \left|\frac{\mathbf{P}_2(0)}{m_2 N_2} - \frac{\mathbf{P}_1(0)}{m_1 N_1}\right|^2\\
&\quad + \nu_{12}\left(E_1(0) + E_2(0) - \frac{1}{2}\frac{|\mathbf{P}_1(0) + \mathbf{P}_2(0)|^2}{m_1 N_1 + m_2 N_2}\right) e^{-\nu_{12} t}\\
&\qquad \cdot \int_0^t e^{\nu_{12} s}\left[\frac{\frac{m_1^{3/2}}{N_1} H_{\tau_1}(c_{12}(s)) - \frac{m_2^{3/2}}{N_2} H_{\tau_2}(c_{21}(s))}{m_1^{3/2} H_{\tau_1}(c_{12}(s)) + m_2^{3/2} H_{\tau_2}(c_{21}(s))}\right] \mathrm{d}s.
\end{aligned}$$

Chapter 3

Numerical schemes

The field of numerical mathematics was eventually born when computers have been invented and the computational power was further developed. New issues arise when a mathematical model is translated to computer language because analytical properties shall be reflected at the numerical level. Clever schemes are developed in order to avoid compromises between small numerical errors and efficient simulations.

We give a short introduction into numerical schemes regarding kinetic equations in Section 3.1. In Section 3.2, we present the basic idea of our scheme for a general multi-species BGK-type equation. Subsequently, we explain the discretization techniques for the corresponding time, space and velocity variables in Sections 3.3–3.5.

3.1 Fundamentals

Kinetic equations depend on three independent variables: time, space and (microscopic) velocity. Each of them needs to be discretized. In this section, we give an overview over existing discretization methods. We start with the discretization in time in Section 3.1.1 because this is the main focus in this thesis. Then we give a short introduction into space discretization techniques in Section 3.1.2. As we will only consider one-dimensional test cases for reasons of computational costs, we also stick to one dimension in the introduction. In Section 3.1.3, we consider the discretization of the three-dimensional velocity space.

3.1.1 Time discretization

We first present methods how to numerically solve an initial value problem for the general autonomous ordinary differential equation (ODE)

$$\begin{aligned} f'(t) &= G(t, f(t)), \\ f(0) &= f^0 \end{aligned} \tag{3.1}$$

with given $G : \mathbb{R} \times \mathbb{R} \to \mathbb{R}$ and $f^0 \in \mathbb{R}$. The exact solution of (3.1) is approximated by $f^\ell \approx f(t_\ell)$ where $t_\ell = \ell \cdot \Delta t$ with $\ell \in \mathbb{N}_0$ and a problem dependent $\Delta t > 0$.

We focus on Runge-Kutta (RK) methods and refer to [HV03].

Runge-Kutta methods

We consider the ODE in (3.1). Generally speaking, we want to calculate $f^{\ell+1}$ from a given f^ℓ. Using RK schemes, intermediate values $f^{[i]}$ are computed, and the next time step $f^{\ell+1}$ is then given by a linear combination of these values. The auxiliary approximations $f^{[i]}$ are called *stages* where $i = 1, \dots, s$. A general form of a RK scheme is given in the following definition.

Definition 3.1.1 (Runge-Kutta method). *Given $A = (a_{ij}) \in \mathbb{R}^{s \times s}$, $\mathbf{b}, \mathbf{c} \in \mathbb{R}^s$ and f^ℓ. The update $f^{\ell+1}$ for* (3.1) *is given by*

$$f^{\ell+1} = f^\ell + \Delta t \sum_{i=1}^{s} b_i G(t_\ell + c_i \Delta t, f^{[i]})$$

with the stage values

$$f^{[i]} = f^\ell + \Delta t \sum_{j=1}^{s} a_{ij} G(t_\ell + c_j \Delta t, f^{[j]}), \quad i = 1, \dots, s.$$

This is often written in a compact way according to [But87] via the Butcher tableau

$$\begin{array}{c|c} \mathbf{c} & A \\ \hline & \mathbf{b} \end{array}.$$

The properties of any RK scheme are determined by A, $\mathbf{b}$ and $\mathbf{c}$. An obvious classification is given by the following definition.

Definition 3.1.2 (Explicit and implicit). *A RK method is called explicit if $a_{ij} = 0$ for $j \geq i$ because all stage values can be computed one after another by explicit relations.*

Otherwise, the method is called implicit. In these cases, the stage values are only given by a system of algebraic relations. If $a_{ij} = 0$ for $j > i$, the scheme is called a diagonally implicit Runge-Kutta (DIRK) scheme.

Let f_e be the exact solution to (3.1). Substituting $f_e(t_\ell)$ into the scheme leads to a value $f_e^{\ell+1}$ which should be approximately $f_e(t_{\ell+1})$ because the method approximates the differential equation. If the *local error* satisfies

$$f_e(t_{\ell+1}) - f_e^{\ell+1} = \mathcal{O}(\Delta t^{p+1}),$$

the method is called *consistent of order p* (provided that G is sufficiently differentiable). Concerning such order of accuracy, so-called *order conditions* can be derived. For first-order schemes, it is required that $\sum_i b_i = 1$. Second-order schemes can be achieved provided that the equality $\mathbf{b}^\top \mathbf{c} = \frac{1}{2}$ is fulfilled. The higher the order, the more conditions on A, $\mathbf{b}$ and $\mathbf{c}$ need to be satisfied. This also results in more and more necessary stages. The minimal number of required stages for a method of order p is determined by *order barriers* [But16]. For schemes of order higher than 5, the number of required stages increases faster than the order of the scheme.

We assume that the method is stable (which is addressed subsequently). Using above notation and starting at the given initial value $f_e(0) = f^0$, the local errors are added up such that the *global error* at time $t = t_L$ is

$$f_e(t_L) - f^L = \mathcal{O}(\Delta t^p).$$

This means that the numerical solution converges to the exact solution for $\Delta t \to 0$. In other words, the method will be convergent with order p when applied to a smooth ODE. However, the order may be reduced for stiff ODEs. Further details on these topics can be found e.g. in [HV03].

A numerical method is called *stable* if small local errors lead to only small global errors. In order to examine the stability of a scheme, the method is applied to the scalar, complex test equation

$$f'(t) = \lambda f(t).$$

One is interested in the stability function R which is determined by the recursion

$$f^{\ell+1} = R(z) f^{\ell}$$

with $z = \lambda \Delta t$. More precisely, one seeks the stability regions

$$S = \{z \in \mathbb{C} \, : \, |R(z)| \leq 1\}.$$

We state the following notions of stability for ODE methods.

Definition 3.1.3 (Stability notions). *An ODE method is called A-stable*[1] *if* $S \supseteq \{z \in \mathbb{C} : \Re(z) \leq 0\}$. *It is said to be strongly A-stable if additionally* $\lim_{z\to\infty} |R(z)| < 1$. *An ODE method is called L-stable if it is A-stable and in addition* $\lim_{z\to\infty} |R(z)| = 0$.

Accordingly, A-stable and L-stable schemes are unconditionally stable which means that they are stable without any condition on the step size.

Remark 3.1.4 (Methods for PDEs). *Consistency, stability and convergence are fundamental requirements for any numerical scheme and become even more challenging if more variables are encountered. Regarding stability, we only mention the notion of total-variation-stability, i.e. the total variation of the numerical solution remains bounded. See e.g. [LeV02].*

We are especially interested in conservation laws. The following is given in [HV03].

Lemma 3.1.5 (Conservation laws for RK schemes). *Let* w *be a given weight function,* $w \not\equiv 0$. *Assume*

$$w(G(t, f)) = 0$$

for all $t \geq 0$ *and* $f \in \mathbb{R}$. *This corresponds to the conservation law that for any solution* $f(t)$ *the quantity* $w(f(t))$ *is constant in time. Then any RK method in Definition 3.1.1 preserves this conservation property; that is*

$$w(f^{\ell+1}) = w(f^{\ell})$$

for all ℓ.

---

1 This notion was introduced in [Dah63].

For later reference, we list several RK schemes.

Definition 3.1.6 (List of RK schemes). *Explicit RK schemes are*

1. *the first-order Forward Euler method*

$$f^{\ell+1} = f^\ell + \Delta t\, G(t_\ell, f^\ell), \tag{3.2}$$

respective

$$\begin{array}{c|c} 0 & 0 \\ \hline & 1 \end{array}. \tag{3.3}$$

2. *the second-order Heun's method*

$$f^{\ell+1} = f^\ell + \frac{\Delta t}{2} G(t_\ell, f^\ell) + \frac{\Delta t}{2} G(t_\ell + \Delta t, f^\ell + \Delta t\, G(t_\ell, f^\ell)), \tag{3.4}$$

respective

$$\begin{array}{c|cc} 0 & & \\ 1 & 1 & \\ \hline & \frac{1}{2} & \frac{1}{2} \end{array}. \tag{3.5}$$

3. *the second-order method ARS-ex [ARS97]*

$$\begin{aligned} f^{[1]} &= f^\ell + \gamma \Delta t\, G(t_\ell + \gamma \Delta t, f^\ell), \\ f^{[2]} &= f^\ell + \delta \Delta t\, G(t_\ell + \Delta t, f^\ell) + (1-\delta)\Delta t\, G(t_\ell + \Delta t, f^{[1]}), \\ f^{\ell+1} &= f^{[2]}, \end{aligned} \tag{3.6}$$

respective

$$\begin{array}{c|ccc} 0 & & & \\ \gamma & \gamma & & \\ 1 & \delta & 1-\delta & 0 \\ \hline & \delta & 1-\delta & 0 \end{array} \tag{3.7}$$

with $\delta = 1 - \frac{1}{2\gamma}$ *and* $\gamma = 1 - \frac{\sqrt{2}}{2}$.

Implicit RK schemes are

1. *the first-order, L-stable Backward Euler method*

$$f^{\ell+1} = f^\ell + \Delta t\, G(t_{\ell+1}, f^{\ell+1}), \tag{3.8}$$

respective

$$\begin{array}{c|c} 1 & 1 \\ \hline & 1 \end{array}. \tag{3.9}$$

2. *the second-order (implicit) and A-stable trapezoidal method*

$$f^{\ell+1} = f^{\ell} + \frac{\Delta t}{2} G(t_\ell, f^\ell) + \frac{\Delta t}{2} G(t_{\ell+1}, f^{\ell+1}), \tag{3.10}$$

respective

$$\begin{array}{c|cc} 0 & & \\ 1 & \frac{1}{2} & \frac{1}{2} \\ \hline & \frac{1}{2} & \frac{1}{2} \end{array} . \tag{3.11}$$

3. *the second-order method ARS-im [ARS97]*

$$\begin{aligned} f^{[1]} &= f^\ell + \gamma\Delta t\, G(t_\ell + \gamma\Delta t, f^{[1]}), \\ f^{[2]} &= f^\ell + (1-\gamma)\Delta t\, G(t_\ell + \Delta t, f^{[1]}) + \gamma\Delta t\, G(t_\ell + \Delta t, f^{[2]}), \\ f^{\ell+1} &= f^{[2]}, \end{aligned} \tag{3.12}$$

respective

$$\begin{array}{c|ccc} 0 & & & \\ \gamma & 0 & \gamma & \\ 1 & 0 & 1-\gamma & \gamma \\ \hline & 0 & 1-\gamma & \gamma \end{array} \tag{3.13}$$

with $\gamma = 1 - \frac{\sqrt{2}}{2}$.

More methods We want to mention two additional classes of schemes because they are also frequently used for kinetic equations.

For an update according to RK schemes, only the previous approximation is needed. This is in contrast to *multi-step methods* where $k \geq 1$ preceding approximations are used for the update. Such methods are defined by

$$\sum_{j=0}^{k} \alpha_j f^{\ell+j} = \sum_{j=0}^{k} \beta_j G(t_{\ell+j}, f^{\ell+j})$$

where the coefficients α_j and β_j depend on the step size Δt and determine the method. The main advantage of these methods is that any update requires only one new evaluation. For RK schemes, every stage corresponds to a new evaluation. Hence, multi-step methods are often used for reasons of efficiency. However, there are two main drawbacks for multi-step methods. Firstly, a change in the step size leads to a change in the coefficients α_j and β_j. Secondly, there is only one value given at initial time. The method yet needs k starting values $f^0, \dots, f^{k-1}$ which therefore need to be calculated by different approaches. [HV03]

An example for multi-step schemes in the context of kinetic equations is provided in [DP17].

Another approach is to use exponential time integrators. Linear or constant parts of the initial value problem can be integrated exactly in time which may help for stiff equations. In the context of BGK equations, we refer to [DP11, DP14] and [Xu01, XCX21, SY08].

Having introduced discretization methods for (3.1), we come to more elaborated equations and schemes.

Implicit-explicit schemes

We consider the generic ODE

$$\partial_t f + \mathcal{T}(f) = \mathcal{R}(f) \tag{3.14}$$

for a distribution function f. In the context of kinetic equations, $\mathcal{T}$ denotes the transport operator, and $\mathcal{R}$ refers to the interaction/relaxation operator. In order to discretize (3.14) in time, both operators and the time derivative need to be discretized. The solution is approximated by $f^\ell \approx f(t_\ell)$.

The structure of the operators indicates a particular treatment. In our case, the transport operator $\mathcal{T}$ is usually discretized explicitly introducing a Courant-Friedrichs-Lewy (CFL) condition which is discussed later in Sections 3.1.2 and 3.4. The relaxation operator $\mathcal{R}$ can also be discretized explicitly. However, for large collision frequencies this results in severe restrictions on the time step size. So often it is preferred to use a more difficult implicit time discretization. Thanks to the special structure of the BGK operators, the implication is comparably easy manageable and the equation stays explicitly solvable [PP07, FJ10]. In Section 3.3.5, we present our general implicit solver for the relaxation operator.

Definition 3.1.7 (IMEX schemes). *If one equation is discretized by a suitable mixture of implicit and explicit methods, one refers to IMEX schemes.*

In the following, we present how IMEX schemes can be constructed and refer to [DP14].

Splitting schemes Operator splitting is a classic and widely used approach, in the kinetic context e.g. in [CP91, HHM17b]. For these methods, the ODE (3.14) is solved in $[t_\ell, t_{\ell+1}]$ as a sequence of steps where only one operator is taken into account during one step. To first order, we first solve the space homogeneous relaxation problem

$$\begin{aligned} \partial_t f^* &= \mathcal{R}(f^*), \\ f^*(t_\ell) &= f^\ell \end{aligned}$$

in $[t_\ell, t_{\ell+1}]$, and for convenience, we write $f^*(t_{\ell+1}) = \mathcal{R}_{\Delta t}(f^\ell)$. Afterwards, we consider the transport step

$$\begin{aligned} \partial_t f + \mathcal{T}(f) &= 0 \\ f(t_\ell) &= f^*(t_{\ell+1}) \end{aligned}$$

in $[t_\ell, t_{\ell+1}]$, and in short $f(t_{\ell+1}) = \mathcal{T}_{\Delta t}(f^*(t_{\ell+1}))$. We summarize this in the following definition.

Definition 3.1.8 (First-order splitting). *Given f^ℓ. The update $f^{\ell+1}$ is achieved by the first-order splitting method*

$$\begin{aligned} f^* &= \mathcal{R}_{\Delta t}(f^\ell), \\ f^{\ell+1} &= \mathcal{T}_{\Delta t}(f^*) = \mathcal{T}_{\Delta t}(\mathcal{R}_{\Delta t}(f^\ell)). \end{aligned}$$

In practice, the splitting is stable provided that every step is stable [HV03]. But additional errors may be introduced by the splitting approach. Only in the case that the operators commute, the splitting is exact [HV03]. Hence, the order of the discretization techniques for the operators is important, as well. It is sensible to perform the operation which acts on shorter time scales at first.

A symmetry in the splitting may yield a better accuracy. Actually, a second-order splitting scheme is given by the well-known *Strang splitting* [Str68].

Definition 3.1.9 (Strang splitting). *Given f^ℓ. The update $f^{\ell+1}$ is achieved by the second-order splitting method*

$$\begin{aligned} f^* &= \mathcal{T}_{\Delta t/2}(f^\ell), \\ f^{**} &= \mathcal{R}_{\Delta t}(f^*) = \mathcal{R}_{\Delta t}(\mathcal{T}_{\Delta t/2}(f^\ell)), \\ f^{\ell+1} &= \mathcal{T}_{\Delta t/2}(f^{**}) = \mathcal{T}_{\Delta t/2}(\mathcal{R}_{\Delta t}(\mathcal{T}_{\Delta t/2}(f^\ell))). \end{aligned}$$

There exist also higher-order splitting schemes [HV03, Chapter IV]. However, we focus on schemes up to second order in this thesis.

The advantage of splitting schemes is obvious: Already existing methods for the individual operators can be glued together. In the special case of kinetic equations, even more benefits exist. The relaxation operator only acts on $\mathbf{v}$, whereas the transport operator acts on $\mathbf{x}$. So the splitting simplifies to design schemes with appropriate properties because the methods for each operator can focus on a reduced number of variables.

Unfortunately, one fundamental drawback is that splitting schemes may suffer from order reduction in the fluid limit [Jin95]. This issue can be overcome by different approaches.

IMEX Runge-Kutta schemes IMEX RK schemes were originally developed in [ARS97] for parabolic PDE. In [Jin95, PR05], they have been extended to hyperbolic systems with relaxation terms. Examples concerning our purposes can be found in [PR05, PP07], and we use such techniques in [HHK$^+$22, BPW22].

Definition 3.1.10 (IMEX RK schemes). *We consider an explicit RK scheme with* $\hat{A} = (\hat{a}_{ij}) \in \mathbb{R}^{\hat{s}\times\hat{s}}$, $\hat{\mathbf{b}}, \hat{\mathbf{c}} \in \mathbb{R}^{\hat{s}}$ *and an implicit RK scheme with* $A = (a_{ij}) \in \mathbb{R}^{s\times s}$, $\mathbf{b}, \mathbf{c} \in \mathbb{R}^{s}$. *Given* f^ℓ. *A standard implicit-explicit Runge-Kutta scheme applied to* (3.14) *can be written as*

$$f^{[i]} = f^\ell - \Delta t \sum_{j=1}^{i-1} \hat{a}_{ij} \mathcal{T}(f^{[j]}) + \Delta t \sum_{j=1}^{s} a_{ij} \mathcal{R}(f^{[j]})$$

$$f^{\ell+1} = f^\ell - \Delta t \sum_{j=1}^{i-1} \hat{b}_i \mathcal{T}(f^{[j]}) + \Delta t \sum_{j=1}^{s} b_i \mathcal{R}(f^{[j]})$$

with the stage values $f^{[i]}$ *and the numerical solution* $f^{\ell+1}$.

Additional to suitable order conditions, the coefficients of both the implicit and the explicit RK scheme need to satisfy coupling conditions [ARS97, KC03, PR05]. A first-order method is given by the coupling of the schemes (3.8) and (3.2), whereas the coupling of (3.12) and (3.6) yields a second-order, L-stable and globally stiffly accurate (GSA) IMEX RK scheme [ARS97]. The notion of GSA schemes is introduced in Definition 3.1.11 below.

In [BPR13, DP14], IMEX RK schemes are characterized and corresponding properties are proven. We only give the following notion.

Definition 3.1.11 (GSA). *An IMEX RK scheme is called globally stiffly accurate if the corresponding DIRK method is stiffly accurate; that is*

$$a_{si} = b_i, \quad i = 1, \dots, s;$$

and if, in addition, the explicit method satisfies

$$\hat{a}_{\hat{s}i} = \hat{b}_i, \quad i = 1, \dots, \hat{s}.$$

According to the previous definition, the numerical solution coincides with the last stage value for GSA schemes. This property is important for the accuracy of the method in the limit where the collision frequencies become infinite.

Structure-preserving properties

In this section, we take a deeper look at two structure-preserving properties, namely AP and strong-stability-preserving (SSP) schemes.

Asymptotic-preserving schemes AP schemes play an essential role in the transition of different regimes, e.g. when passing from kinetic to fluid descriptions. In this case, a great number of collisions occur making the interaction part stiff, but the collisional forces are not strong enough to drive the kinetic system to a fluid description yet. In order to deal with such multi-scale problems, unconditionally stable numerical methods have been developed which succeed in capturing the correct asymptotic behavior while avoiding severe restriction on the time step. These schemes are called AP [DP14]. To be more precise, we give the following definition.

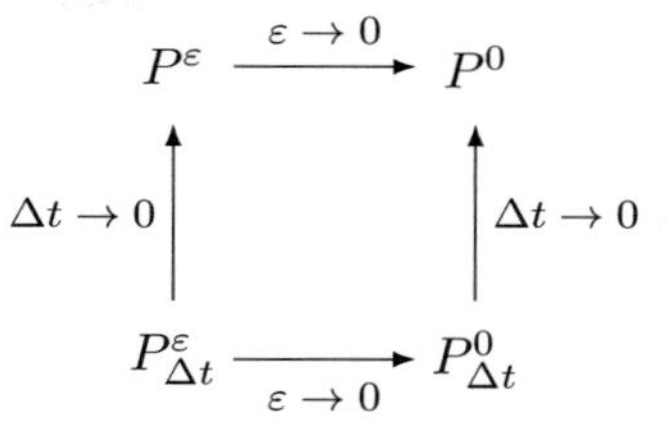

Figure 3.1: A kinetic problem P^ε and its fluid limit P^0 are discretized by $P^\varepsilon_{\Delta t}$ and $P^0_{\Delta t}$. If the diagram commutes, that is if $P^\varepsilon_{\Delta t}$ is a consistent and stable approximation of P^0 as $\varepsilon \to 0$ for a fixed Δt, the method is called AP.

Definition 3.1.12 (AP). *Let P^ε be a kinetic problem with fluid limit P^0 for $\varepsilon \to 0$. A consistent and stable time discretization $P^\varepsilon_{\Delta t}$ of step size Δt is called asymptotic preserving if it becomes a consistent and stable time discretization $P^0_{\Delta t}$ for the reduced fluid equation in the limit $\varepsilon \to 0$ and for fixed Δt, see Figure 3.1.*

The scaling parameter can be associated with the Knudsen number which appears e.g. in the nondimensional Boltzmann equation [Str05]. For BGK equations, $\varepsilon \to 0$ corresponds to enormous collision frequencies $\nu \to \infty$.

Using AP schemes, the correct equilibrium solutions are preserved while remaining an efficient scheme [Pup19, HJL17, BIP15, DP14, FJ10, BLM08, PP07]. In this context, splitting methods should be treated with care because, in the hydrodynamic limit, the (second-order) Strang splitting reduces actually to a first-order approximation of the equilibrium equation as shown in [Jin95].

Strong-stability-preserving schemes For this kind of schemes, we need to anticipate parts of the space discretization. Considering a PDE, it is a common practice to first discretize the space variables which results in a system of ODEs in the time variable. ODE solvers can be applied provided that an adequate stability criterion is satisfied. For hyperbolic equations, a corresponding class of (high-order) time discretization methods is called *total-variation-diminishing* (TVD), introduced in [Shu88, SO88]. However, in [GST01] the authors explain that the notion *strong-stability-preserving* (SSP) is more suitable for this class because the essential part of these methods is not restricted to the total variation norm. The construction of such schemes often relies on convexity arguments which hold for any norm. The property is given in the following definition.

Definition 3.1.13 (SSP). *A scheme is called strong-stability-preserving if*

$$||f^{\ell+1}|| \leq ||f^\ell||$$

for any norm $||\cdot||$.

If an equation possesses an entropy decay structure, this one is preserved at the discrete level as a consequence of the SSP property. Moreover, the positivity of an IMEX RK scheme follows from the above property. More generally speaking, an

SSP scheme preserves any convex property which makes SSP schemes attractive. However, in [GST01] it was shown that unconditionally SSP implicit schemes are at most first-order accurate. As the positivity of the distribution functions indeed is required, further developments are needed for kinetic equations.

For the standard BGK equation, such an extension is given in [HSZ18] where an additional correction step to a preceded IMEX RK method guarantees the positivity of the scheme. In [GGHS22], the SSP property for high-order schemes can be achieved by using multi-derivative IMEX RK methods. The schemes in [HSZ18, GGHS22] ask for several requirements concerning the individual positivity preservation of the transport and the interaction operator, respectively, and some dissipative behavior of the collision operator. To the best of our knowledge, one of those cannot be proven for velocity-dependent collision frequencies such that we cannot apply these schemes to our equations (2.69). This also means that we need to take care of the positivity preservation by additional considerations, see Section 4.2.

In the following section, we concern ourselves with existing methods for the discretization of the space variable.

3.1.2 Space discretization

There can be found many different approaches for the space discretization in the literature as the same transport term is incorporated by many different kinetic equations. The most common techniques for the space discretization are finite difference (FD), finite element (FEM) and finite volume (FV) schemes. For FD methods, the derivatives are approximated by finite differences, and discrete point values are evolved in time. This results in a system of linear equations. For FEM schemes, the considered matter is divided into a finite number of elements. The evolution of these elements is described in time based on the forces acting on and between them. Whereas the space is discretized into control volumes for FV schemes, and occuring fluxes between such volumes are approximated.

The transport term being hyperbolic, a FV discretization is often used, preserving the conservation properties by construction. That is why we will use this approach as well. In the following, we give a short introduction into FV schemes. For reasons of computational costs, we will only consider one-dimensional settings in space (but we will stick to three dimensions in velocity, see Remark 3.2.1). Therefore, we restrict the following introduction to one dimension. For extensions to higher dimension and for more information, we refer to [LeV02, Tor09, Ber21].

Finite Volume schemes

A convenient approach for conservation laws is a FV method. Let us consider the following one-dimensional balance law in the spatial domain $I \subset \mathbb{R}$

$$\partial_t f(x,t) + \partial_x F(f(x,t)) = \mathcal{R}(f) \quad \text{for } x \in I, t > 0 \tag{3.15}$$

$$f(x,0) = f^0(x) \quad \text{for } x \in I, \tag{3.16}$$

where F denotes the flux function. In kinetic equations, it is $F(f) = vf$.

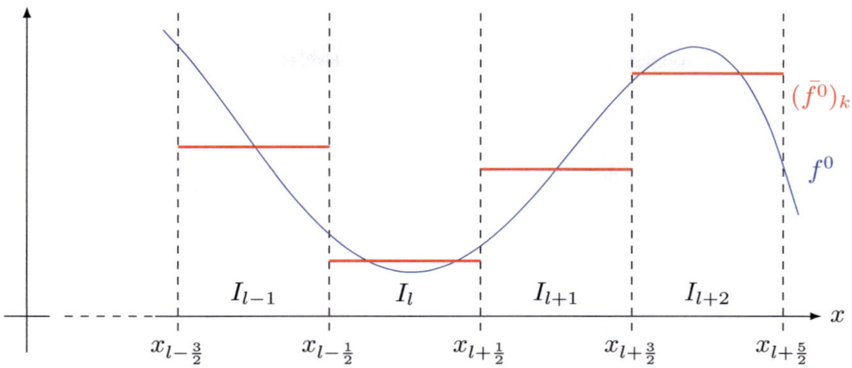

Figure 3.2: The initial data f^0 is given and averaged in each cell I_k. The resulting function is piecewise constant with values $(\bar{f^0})_k$ for $x \in I_k$.

We divide the domain in cells $I_k = [x_k - \frac{\Delta x}{2}, x_k + \frac{\Delta x}{2}]$ for $k \in \{0, \dots, K\}$ and an appropriately chosen $\Delta x > 0$. For simplicity, we use equidistant cells. In every cell, the initial data is averaged

$$(\bar{f^0})_k := \frac{1}{\Delta x} \int_{I_k} f^0(x)\,\mathrm{d}x$$

resulting in a piecewise constant function

$$\bar{f}(x,0) = (\bar{f^0})_k \quad \text{for } x \in I_k,$$

illustrated in Figure 3.2. We now look at the modified initial value problem

$$\partial_t \bar{f} + \partial_x F(\bar{f}) = \mathcal{R}(\bar{f}) \quad \text{for } x \in I, t > 0 \tag{3.17}$$

$$\bar{f}(x,0) = (\bar{f^0})_k \quad \text{for } x \in I_k. \tag{3.18}$$

These equations describe a series of *Riemann problems*; that is, a Cauchy problem where the initial condition consists of two constant states seperated by a discontinuity. We integrate (3.17) in space and obtain an ODE which can be solved according to Section 3.1.1. The equation in conservative form reads

$$\partial_t \bar{f}_k = \bar{f}_k - \frac{1}{\Delta x}\left(\mathscr{F}_{k+\frac{1}{2}} - \mathscr{F}_{k-\frac{1}{2}}\right) + \mathcal{R}_k \tag{3.19}$$

where

$$\bar{f}_k = \frac{1}{\Delta x} \int_{I_k} f(x,t)\,\mathrm{d}x$$

are cell averages. Using the set of cell averages $\{\bar{f}_k\}$, f can be reconstructed inside a cell I_k resulting in a polynomial $p_k(x) \approx f(x)$ for $x \in I_k$ [Tor09]. Often, the polynomials p_k and p_{k+1} are used to determine the flux across the interace $x = x_{k+\frac{1}{2}}$. In general, we approximate the flux at the interface $x = x_{k+\frac{1}{2}}$ as a function of adjacent cell averages

$$\mathscr{F}_{k+\frac{1}{2}} = \mathscr{F}(\bar{f}_{k-l}, \ldots, \bar{f}_{k+r}) \tag{3.20}$$

with l and r two nonnegative integers. Moreover, we define the space integral average

$$\mathcal{R}_k = \frac{1}{\Delta x} \int_{I_k} \mathcal{R}(f(x,t))\, \mathrm{d}x. \tag{3.21}$$

The formulae (3.20) and (3.21) need to be approximated appropriately leading to *numerical fluxes* and *numerical sources*, respectively. The construction of numerical fluxes is a central task for the numerical method. There exist approaches where the Riemann problems are solved exactly such as *Godunov's method* applied to the Euler equations. Alternatively, approximations are used which result in different numerical fluxes. In any case, the numerical fluxes are supposed to satisfy some properties.

Definition 3.1.14 (Consistent numerical flux). *The numerical flux $\mathscr{F}$ is called consistent with the physical flux F if $\mathscr{F}(g, \ldots, g) = F(g)$.*

A sufficient criterion for consistency is to have a Lipschitz continuous function $\mathscr{F}$ in each variable. A consistent numerical flux ensures consistency of the semi-discrete scheme (3.19) with the hyperbolic conservation law (3.15). If the numerical method is additionally stable, the scheme converges under grid refinement. This is often summarized in the *fundamental theorem of numerical methods for PDEs*. For further information, including different notions of stability, we refer to [LeV02].

In this context, we want to mention the *Lax-Wendroff theorem* [LW60] which guarantees that the approximate solution of a consistent and conservative scheme for a hyperbolic conservation law with convergence under grid refinement converges in this case towards a weak solution of the conservation law.

A further issue are the boundaries, either physically given by the spatial domain or artificially set for computational reasons. For *periodic* boundaries, the cells I_K and I_0 are thought of to be connected. But also *open* and *fixed* boundaries can be used. Any boundary behavior results in a special treatment of the numerical fluxes $\mathscr{F}_{K+\frac{1}{2}}$ and $\mathscr{F}_{-\frac{1}{2}}$.

High-resolutions can be obtained by elaborated methods, e.g. by weighted essentially non-oscillatory (WENO) schemes. And discontinuous Galerkin (DG) schemes can also be applied which is a combination of the FEM and FV framework. For orders higher than two, one has to be especially careful with the corresponding formulation of the numerical source $\mathcal{R}_k$. [Mie00, PP07, HJL17, BCS11, CGP12]

More approaches

Especially for the transport term, semi-Lagrangian (SL)[2] methods have been developed which acknowledge its special and quite easy structure. The full equation is then often solved by splitting methods [DP14]. For the SL approach, the characteristics, being straight lines (see Section 2.1.2), are followed exactly. This requires an interpolation for the evaluation of the corresponding foot point, illustrated in Figure 3.3. A benefit of these schemes is that no stability conditions are needed. Unfortunately, SL methods are not conservative in general. Nevertheless, they can be kept conservative also for higher orders by conservative reconstructions or corrections [SRBG99, CMS10, DP14, CBRY21, QS11].

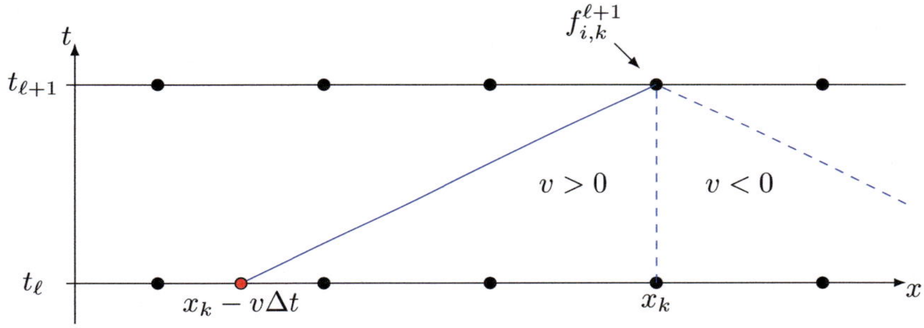

Figure 3.3: We illustrate the SL approach for the transport term. To determine $f_{i,k}^{\ell+1}$, the characteristic curve is followed for $v > 0$. Since the foot point $x_k - v\Delta t$ does not hit a node of the grid, an interpolation is needed.

Based on discrete-velocity models (see Section 3.1.3) and SL methods, efficient schemes can be developed [DL13]. Here, the distribution function needs not to be reconstructed at each time step (in contrast to standard SL schemes) which then accelerates numerical computations.

We also want to mention Particle in Cell (PIC) methods. These are the most used methods for the Vlasov equation which approximate the plasma by a finite number of macro-particles [FSB01]. In [FSB01], the authors shortly discuss different methods for the Vlasov equation and then introduce their positive and flux conservative method (PFC).

We have given an overview over existing techniques for discretizing the space and time variable. In order to obtain a fully-discrete scheme, the velocity variable needs to be discretized as well.

---

2 The notion reminds of the description in Lagrangian coordinates in contrast to Eulerian coordinates. For these, see [TM05].

3.1.3 Velocity discretization

Considering the microscopic velocities as independent variables, poses several challenges. In the same time, it introduces more degrees of freedom which makes it possible to capture more phenomena at smaller scales.

Historically, discrete-velocity methods aimed to study a rarefied gas qualitatively [DP14]. A relatively new approach are the *Lattice-Boltzmann methods* which can be seen as schemes for a discrete-velocity Boltzmann equation. This has been successfull in simulating hydrodynamic systems up to the Navier-Stokes level with possible complex boundaries. Accordingly, discrete-velocity BGK models have been developed. See [Bue96, Mie00, Mie01, YH09], as well as [DP14, Chapter 4] and references therein.

Regarding computational costs, kinetic equations suffer from the high dimensionality. So it is recommendable to use coarse grids [Pup19] whenever possible. However, a coarse velocity grid amplifies the challenges regarding errors in the macroscopic quantities which makes it necessary that given conservation laws are fulfilled at the discrete level as well.

In [GT09], the authors propose a constrained L^2-projection in order to preserve the moment conservation properties. A profound study of discrete moments, a discrete entropy and the corresponding discrete Maxwellians in a BGK model is provided in [Mie00]. There, basic estimates on and effects to the mean velocities and temperatures can be found when microscopic velocities are discretized and truncated. We shortly summarize some of the statements in the following. For reference, see [Mie00, DP14].

A discrete-velocity model for the BGK equation

Let us consider one space and one velocity dimension for sake of simplicity in notation. The BGK equation for one species with mass m (and constant in v collision frequency) reads

$$\partial_t f + v\,\partial_x f = \nu(\mathcal{M}[f] - f).$$

We introduce a discrete velocity grid

$$\{v_q \in \mathbb{R}, \quad q \in V\}$$

with $V \subset \mathbb{Z}$ being a finite subset and $\Delta v > 0$ such that a node of the mesh can be written as

$$v_q = q\Delta v.$$

We are interested in the evolution of the vector $\mathbf{f} = (f_q(x,t))_q \approx (f(x,v_q,t))_q$ which is described by a system of $|V|$ partial differential equations[3]

$$\partial_t f_q + v_q\,\partial_x f_q = \nu_V(\mathcal{M}_q[\mathbf{f}] - f_q),$$

[3] For a fully-discrete scheme, an additional discretization in space and time is needed.

where the discrete Maxwellian $\mathcal{M}_q[\mathbf{f}]$ is not defined yet. The discrete collision frequency $\nu_V = \nu(n_V, T_V)$ is determined by the discrete fluid quantities[4] which can be calculated by discrete sums. We use the midpoint rule; that is

$$\begin{aligned} n_V &= \sum_{q \in V} f_q \, \Delta v, \\ n_V u_V &= \sum_{q \in V} v_q f_q \, \Delta v, \\ n_V T_V &= \sum_{q \in V} m(v_q - u_V) f_q \, \Delta v. \end{aligned} \tag{3.22}$$

We define the discrete entropy

$$\mathcal{H}_V[\mathbf{f}] = \sum_{q \in V} f_q \log(f_q) \, \Delta v.$$

We want to find $\mathcal{M}_q[\mathbf{f}]$ such that the discrete conservation laws

$$\begin{aligned} \sum_{q \in V} (\mathcal{M}_q - f_q) \, \Delta v = 0, \\ \sum_{q \in V} v_q (\mathcal{M}_q - f_q) \, \Delta v = 0, \\ \sum_{q \in V} m|v_q - u_V|^2 (\mathcal{M}_q - f_q) \, \Delta v = 0 \end{aligned} \tag{3.23}$$

are satisfied, as well as a corresponding discrete formulation of the $\mathcal{H}$-Theorem.

Discrete Maxwellians A natural choice seems to be

$$\mathcal{M}_q[\mathbf{f}] = \mathcal{M}[n_V, u_V, T_V, m] = \frac{n_V}{(2\pi T_V/m)^{\frac{1}{2}}} \exp\left(-\frac{m|u_V - v_q|^2}{2T_V} \right). \tag{3.24}$$

However, this formula, in general, does not satisfy neither the conservation properties (3.23) nor the entropy behavior. Instead, the discrete equilibrium $\mathcal{M}_q[\mathbf{f}]$ is defined in the following way.

Definition 3.1.15 (Discrete Maxwellian). *We call*

$$\mathcal{M}_q[\mathbf{f}] = \mathcal{M}_q^{a,b,c} = \exp(a + bv_q + c|v_q|^2), \quad a, b, c \in \mathbb{R} \tag{3.25}$$

a discrete Maxwellian if and only if there exists $\mathbf{f} > 0$ resulting in the discrete moments n_V, u_V and T_V.

---

4 Since we consider only one velocity dimension in this section, the factor in the temperature differs from the factor in the rest of the thesis (where we consider three velocity dimensions).

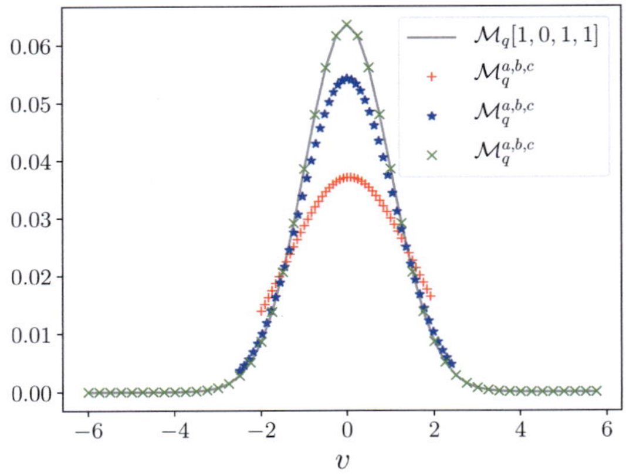

Figure 3.4: The Maxwellian $\mathcal{M}_q[1,0,1,1]$ according to (3.24) is illustrated by the gray line. In comparison, discrete Maxwellians $\mathcal{M}_q^{a,b,c}$ are plotted, all of which share the same discrete moments $n_V = 1$, $u_V = 0$, $T_V = 1$. However, the corresponding parameters a, b, c for these discrete Maxwellians (3.25) are calculated with respect to different velocity grids (in width) resulting in distinct functions.

The necessary and sufficient condition comes from the particular choice of the grid as not all sets of moments may be realizable by the given discrete velocities. Additionally, the velocity grid needs to be large enough. But this last statement is no actual restriction; it is suffcent to take a Cartesian grid with at least two nodes in each direction and at least three nodes in a given direction.

The crucial difference between (3.24) and (3.25) is that the parameters $a, b, c \in \mathbb{R}$ cannot be computed explicitly from the given moments n_V, u_V and T_V. These parameter need to be determined by solving (3.23) via a Newton iteration method, for instance.

More information (including existence, uniqueness and convergence results) and the proofs of the given statements can be found in [Mie00].

We illustrate discrete Maxwellians in Figure 3.4. The Maxwellian $\mathcal{M}_q[1,0,1,1]$ according to (3.24) is provided by the gray line. If we instead force the discrete Maxwellian (3.25) to have the discrete moments $n_V = 1$, $u_V = 0$ and $T_V = 1$, the resulting functions highly depend on the underlying velocity grid. The velocity domain used for the discrete Maxwellian $\mathcal{M}_q^{a,b,c}$ given by the green crosses is large enough such that $\mathcal{M}_q^{ab,c}$ is comparable to $\mathcal{M}_q[1,0,1,1]$. However, for smaller velocity grids, the corresponding $\mathcal{M}_q^{ab,c}$ differ significantly from $\mathcal{M}_q[1,0,1,1]$, e.g. for the blue stars and the red crosses, respectively.

Temperature and mean velocity bounds We already mentioned that the set of moments being realizable by the discrete velocity grid is restricted. This means that, once a velocity grid is chosen, the corresponding discrete-velocity model cannot describe any gas flow. Conversely, for a given flow, the velocity space must be discretized properly in order to obtain a good description.

Proposition 3.1.16 (Bounds for discrete macroscopic quantities). *Let f be a distribution function and the velocity grid given. Then the discrete mean velocity u_V and temperature T_V associated with* $\mathbf{f}$*, defined in* (3.22)*, satisfy*

$$\min_V v_q \leq u_V \leq \max_V v_q, \tag{3.26}$$

$$\min_V |v_q - u|^2 \leq T_V \leq \max_V |v_q - u_V|^2. \tag{3.27}$$

The previous proposition illustrates that the velocity grid must be large enough for a possible resolution of high velocities and high temperatures. Moreover, the temperature is also bounded from below. This means that Δv must be chosen small enough in order to resolve possibly small temperatures as pointed out in [Mie00].

Quadrature

We are interested in macroscopic quantities being represented by integrals over the distribution function with respect to the microscopic velocity. By discretizing the velocity space, the question arises how to discretize integrals. Usually, function values at specific points are weighted and summed up, specified by the corresponding quadrature rule:

$$\int_a^b \omega(v) f(v)\,\mathrm{d}v \approx \sum_{i=1}^n \omega_i f(v_i) \tag{3.28}$$

where $f(v)$ is a given function, $\omega(v)$ is a weight function, ω_i are discrete weights and v_i are the corresponding (possibly unequally spaced) nodes for $i = 1, \ldots, n$.

Gaussian quadrature rules are popular being constructed such that polynomials of degree $2n-1$ or less are maintained exactly. However, the location of the nodes v_i poses difficulties for kinetic equations. We illustrate this for the Gauss-Hermite quadrature which is designed to integrate functions on unbounded intervals of the following form:

$$\int_{\mathbb{R}} v^s f(v) e^{-v^2}\,\mathrm{d}v \approx \sum_{i=1}^n \omega_i v_i^s f(v_i).$$

The weight function $\omega(v)$ is a Maxwellian, and polynomials of degree $2n-1$ are maintained exactly if $2s-1 = n$. As we are especially interested in $s = 0, 1, 2$, a small velocity grid with $n = 3$ sufficed. This method is very suitable for functions being close to Maxwellians. However, the parameters of the Maxwellian (mean velocity, temperature) are unknown and may change in space and time. This makes it hard to define a suitable velocity grid and the corresponding nodes v_i. Accordingly, different quadrature rules are often employed. [Pup19]

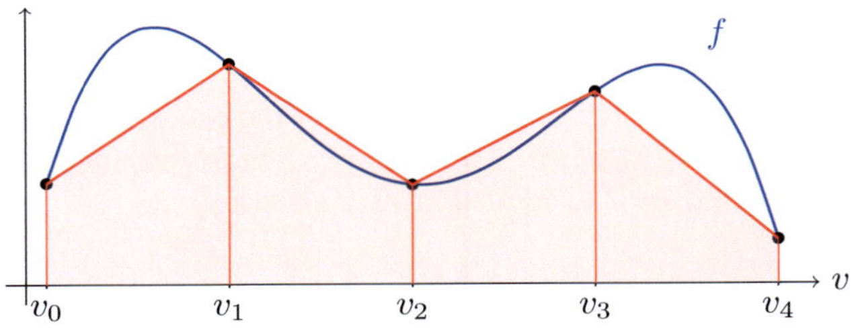

Figure 3.5: Illustration of the composite trapezoidal rule for 4 subintervals. The integral of f is approximated by the area of a trapezoid in each subinterval. The corresponding weights are $\tilde{\omega}_0 = \tilde{\omega}_4 = \frac{1}{2}$ and $\tilde{\omega}_1 = \tilde{\omega}_2 = \tilde{\omega}_3 = 1$.

A simpler approach is to use an equally spaced velocity grid leading to the *Newton-Cotes* formulas. For reference see e.g. [SB02, QSS07]. The weight function $\omega(v) = 1$ in (3.28) is constant, and the discrete weights ω_i can be computed as the integral of Lagrange basis polynomials l_i:

$$\int_a^b f(v)\,\mathrm{d}v \approx \int_a^b \sum_{i=0}^{n} f(v_i) l_i(v)\,\mathrm{d}v = \sum_{i=0}^{n} f(v_i) \underbrace{\int_a^b l_i(v)\,\mathrm{d}v}_{=:\omega_i}.$$

Instead of using high order polynomials for large n, we apply a composite rule; that is, the interval $[a, b]$ is split into smaller subintervals, and the quadrature rule is executed on each subinterval.

For $n = 1$, we obtain the *trapezoidal rule* having spectral accuracy for smooth and periodic functions on a uniform grid. [BIP14] The method's name comes by the fact that the area under the graph of $f(v)$ is approximated by a trapezoid. Hence, the discrete weights are $\omega_i = \frac{b-a}{2}$. For the composite rule, we assume N subintervals and the nodes $v_0, \ldots, v_N$. We obtain

$$\int_{v_0}^{v_N} f(v)\,\mathrm{d}v \approx \sum_{i=0}^{N} f(v_i)\tilde{\omega}_i \Delta v$$

with the weights

$$\tilde{\omega}_i = \begin{cases} 1 & \text{for} \quad i = 1, \ldots, N-1, \\ \frac{1}{2} & \text{else} \end{cases}$$

and $\Delta v = \frac{b-a}{N}$. We illustrate the composite trapezoidal rule in Figure 3.5.

Further remarks on velocity grids

For multi-species equations, the mass ratio between species strongly influences the performance of schemes. For instance, multi-species Boltzmann equations with significant differences in species mass require expensive grid resolutions [MTH+14] because of the direct integration of distribution functions in the Boltzmann operator (2.29). Here comes another advantage of multi-species BGK equations. Since particles of different species only interact with each other through moments, each equation in the BGK model can be discretized on a seperate grid [HHM17b].

Grid adaption might be an important tool when the mean velocities $\mathbf{u}(\mathbf{x}, t)$ cover a wide range and small temperatures are encountered. Accordingly, grid adaption is tackled more and more in the last decade, e.g. in [BM14, BIP14, BCR22, HB13].

Another approach for a considerable reduction of the computational costs is the Chu reduction [Chu65] if there are more degrees of freedom in velocity than in space. Being interested in macroscopic quantities only, the system's dimensionality is reduced by looking at the evolution of appropriate integrals of the distribution functions, but these integrals are no macroscopic quantities yet.

We have presented discretization techniques for the velocity variable. Many concepts can be carried over for quantum models by replacing the velocity with momentum.

We conclude this section by highlighting existing literature on schemes for BGK equations.

3.1.4 Existing schemes for BGK equations

In the literature, numerous approaches for discretizing kinetic equations can be found. We already mentioned in the previous sections which discretization techniques are often used. We now highlight identified schemes for BGK equations, partly given in [PWed].

We start with contributions where the special structure of the BGK model is used in a wider context. In these, the benefit from the reduced computational complexity compared to the Boltzmann equation is crucial. This is useful, for instance, for penalization techniques; in [FJ10, JL13], the BGK equation serves as preconditioner for the numerical solution of the Boltzmann equation. The authors in [DDP11] aim to develop an improved Monte Carlo simulation of the Boltzmann equation. As a first step, they have developed such a method for the BGK equation. Moreover, the BGK approach manages to couple different domains in which the regimes range from equilibrium to very rarefied [AP12].

This gives rise to the observation that simulations of the BGK model itself are of interest in the community. We want to mention two specific approaches for AP schemes here. The *micro-macro decomposition* [CCL12, CKP20] is based on writing the distribution function as a sum of its equilibrium part and the remnant kinetic part. This decomposition results in one kinetic and one macroscopic equation which can be solved by individual and adequate methods. The former equation is solved by a particle method, whereas the fluid part is solved by a standard FV approach.

This discretization technique fits very well to plasma applications. In the core of a plasma, a kinetic description is mandatory. However, next to the wall of a Tokamak, the plasma is close to a fluid such that a hybrid kinetic/fluid description is adequate.

Another approach is the *parity decomposition/AP splitting* [JP00, DP14] where the distribution function is split into an even and an odd parity. This results in a new system of equations with only one time scale where splitting techniques can be applied appropriately.

A totally different method is the *low-rank* approximation. The resulting evolution equations describe the dynamics of the model constraint to the corresponding low-rank manifold which goes with a reduction of the dimensionality of the problem. Hence, dynamical low-rank algorithms provide robust and efficient approximations to several kinetic models, among others to the BGK equation. [Ein19, EHY21]

There also exist contributions using exponential time integrators, often combined with splitting techniques. The idea is to integrate parts of the equations exactly and include modelling aspects in the remnant. This results e.g. in a distinct treatment of collisional and collisionless particles. [DP11, Xu01, XCX21] The latter article additionally takes care of velocity-dependent collision rates.

In [MS04], a numerical scheme for a one-species BGK model with velocity-dependent collision frequency [Str97] is presented. Their time-explicit scheme follows a CFL condition being restrictive if the maximal value of the collision frequency gets large. As time-implicit method, the authors propose to linearize the target function around the ansatz at the current value which results in an efficient scheme for the simulation of steady-state solutions. However, as noted in [Mie00], this approach lacks conservation and entropy properties at the discrete level.

Active contribution in numerics for multi-species BGK models has strongly increased in the last years. At the discrete level, many ideas can be carried over from the single-species schemes. In the following, we give contributions of numerical schemes for multi-species BGK equations, which can be written in the form of the KPP model (2.51).

We already mentioned [CKP20] as example of the micro-macro decomposition approach.

In [HHM17a], the authors are interested in capturing physical transport coefficient. They match the relaxation rates in the space homogeneous case to the Boltzmann ones as described in Section 2.1.4. An extension to space inhomogeneous settings is done in [HHM17b], and they examine the coupling to electric fields as well.

In [BCGR21], the authors compare numerical results for different multi-species BGK models. One of those models is a special case of (2.51).

Numerical schemes for quantum equations are less represented in the literature. Still, several articles on schemes for quantum kinetic equations can be found. For

BGK models, existing schemes are extended which often includes additional root finding algorithms in order to solve algebraic equations [FHJ12, WMZ12, MY12]. In [YH09], a lattice method is provided based on Grad's method. A micro-macro decomposition is performed in [CM14]. Moreover, a relaxation time approximation can be found in [SY08].

Having presented identified contributions concerning schemes for BGK equations, we now shortly summarize existing approaches for the discretization of the relaxation operator.

A straight-forward method is a time-explicit discretization as the target function can be determined directly from given data. By establishing discrete Maxwellians (see Section 3.1.3), the velocity discretization becomes more accurate. [Mie00]

Time-implicit techniques are more complicated because the value of the target function at the next time step is not provided ad-hoc. Indeed, for the standard one-species BGK model, the macroscopic quantities stay constant in time as does the target function. Hence, the implication corresponds to a linear solve [CP91, PP07, RF09, FJ10, DP13]. But for more general models, the implication goes with more difficulties. In some set-ups (e.g. multi-species BGK models with constant collision frequencies), the target functions can be evaluated implicitly with the help of a moment update using an iterative solver [Pup19]. Numerical schemes for quantum models additionally apply root finding algorithms [WMZ12]. Another approach is a linearization of the target function [Mie00, MS04].

We are interested in a time-implicit discretization of the relaxation term. However, the given methods in the literature are not applicable to the multi-species BGK model with velocity-dependent collision frequencies, or these do not meet our requirements. Thus, we develop a new scheme enabling time-implicit target functions while guaranteeing conservation and entropy properties at the discrete level. The main new ingredient is a general implicit solver for the target functions.

3.2 Basic idea of our scheme

We present the basic idea of our scheme for multi-species BGK equations. This approach works for many different settings, but we focus especially on velocity-dependent collision frequencies and quantum models, respectively, which make a special treatment of the target functions necessary. Parts of the following can already be found in [HHK+22, BPW22].

The scheme is a discrete-velocity method which relies on standard time and space discretizations from the literature.

As introduced in Section 3.1.1, we use an IMEX time discretization. Transport terms are treated explicitly, whereas interaction parts are treated implicitly because large collision frequencies make these terms stiff. We follow splitting methods and IMEX RK methods up to second order, but the ideas can be carried over to different discretization techniques as well. The key new ingredient is a solver which enables an implicit treatment of the BGK operator with a broad class of target functions. Mimicking the analytic case, the crucial step is the formulation of a convex entropy minimization problem. The solver is based on a numerical minimization procedure in order to determine the parameters in the target functions. By this construction, conservation and entropy properties are preserved at the discrete level, up to numerical tolerances.

The space discretization relies on a FV method. We assume a slab geometry for which $\partial_{x^{(2)}} f = \partial_{x^{(3)}} f = 0$. So we reduce the physical space dimension to one space dimension and write $x = x^{(1)}$. The numerical fluxes are inspired by [MS04] and preserve the positivity of the distribution functions provided that an adequate CFL condition is fulfilled.

The (microscopic) velocity space remains three-dimensional ($\mathbf{v} = (v^{(1)}, v^{(2)}, v^{(3)})$). A discrete-velocity method for the equations is proposed similar to [Mie00]. The optimization algorithm is performed on an adequate velocity grid.

Remark 3.2.1 (1d3v). *Using one space dimension but three velocity dimension is often abbreviated by* 1d3v. *This set-up is of real interest as illustrated by still air. Even though in the macroscopic framework, there is no movement in the air, molecules do move around in all three directions at the particle level. So it does make sense to take into account three microscopic velocity directions even if the macroscopic movement is zero- or one-dimensional. In other words, we consider only flows with mean velocities* $\mathbf{u} = (u^{(1)}, 0, 0)$.

We give a general formulation for multi-species BGK equations which covers both, the model in Section 2.2.2 with velocity-dependent collision frequencies and the quantum model in Section 2.3. Hence, we introduce an auxiliary variable

$$\mathbf{w} = \begin{cases} \mathbf{v} & \text{for the model in Section 2.2.2 with } \mathbf{v}\text{-dependent collision frequencies,} \\ \mathbf{p} & \text{for the quantum model in Section 2.3.} \end{cases}$$

In the following, we identify the corresponding models by the value of $\mathbf{w}$. We consider two species $i, j \in \{1, 2\}$ for simplicity. But the extension to more species is straight-forward as interactions are assumed to be binary (outlined in Appendix B for 3 species). The system of PDEs reads

$$\begin{aligned} \partial_t f_1 + \mathcal{T}(f_1) &= \mathcal{R}_1(f_1, f_2) \\ \partial_t f_2 + \mathcal{T}(f_2) &= \mathcal{R}_2(f_2, f_1) \end{aligned} \tag{3.29}$$

with the transport operator

$$\mathcal{T}(f_i) = v^{(1)} \partial_x f_i$$

and the relaxation operator

$$\mathcal{R}_i(f_i, f_j) = \nu_{ii}(\mathbf{w})(\mathcal{A}_{ii,\tau_i}(\mathbf{w}) - f_i) + \nu_{ij}(\mathbf{w})(\mathcal{A}_{ij,\tau_i}(\mathbf{w}) - f_i). \tag{3.30}$$

Remark 3.2.2. *We use the velocity* $\mathbf{v} = \frac{\mathbf{p}}{m_i}$ *in the transport operator. However, this causes problems for massless particles like photons (which we neglect in this thesis). In such cases, a more general formulation of the velocity is required, see e.g. [EMV03].*

The collision frequencies depend on the corresponding model; that is

$$\nu_{ij}(\mathbf{w}) = \begin{cases} \nu_{ij}(\mathbf{v}) & \mathbf{w} = \mathbf{v}, \\ 1 & \mathbf{w} = \mathbf{p}. \end{cases}$$

We introduce the notation $\mathbf{a}_i = (\mathrm{a}_i^{(0)}, \mathbf{a}_i^{(1)}, \mathrm{a}_i^{(2)})^\top \in \mathbb{R}^5$ where

$$\mathbf{a}_i(\mathbf{v}) = m_i \begin{pmatrix} 1 \\ \mathbf{v} \\ |\mathbf{v}|^2 \end{pmatrix} \quad \text{and} \quad \mathbf{a}_i(\mathbf{p}) = \begin{pmatrix} 1 \\ \mathbf{p} \\ \frac{|\mathbf{p}|^2}{m_i} \end{pmatrix}.$$

The target functions are given by

$$\mathcal{A}_{ij,\tau_i}(\mathbf{w}) = \frac{1}{\exp(-\mathbf{a}_i(\mathbf{w}) \cdot \boldsymbol{\lambda}_{ij}) + \tau_i(\mathbf{w})} \tag{3.31}$$

with

$$\tau_i(\mathbf{w}) = \begin{cases} 0 & \text{for } \mathbf{w} = \mathbf{v}, \\ 0 & \text{for } \mathbf{w} = \mathbf{p} \text{ and for species } i \text{ being classic particles,} \\ +1 & \text{for } \mathbf{w} = \mathbf{p} \text{ and for species } i \text{ being fermions,} \\ -1 & \text{for } \mathbf{w} = \mathbf{p} \text{ and for species } i \text{ being bosons.} \end{cases}$$

The main task will be to determine the parameters $(i, j = 1, 2;\ i \neq j)$

$$\boldsymbol{\lambda}_{ii} = \begin{pmatrix} \lambda_{ii}^{(0)} \\ \boldsymbol{\lambda}_{ii}^{(1)} \\ \lambda_{ii}^{(2)} \end{pmatrix} \quad \text{and} \quad \boldsymbol{\lambda}_{ij} = \begin{pmatrix} \lambda_{ij}^{(0)} \\ \boldsymbol{\lambda}^{(1)} \\ \lambda^{(2)} \end{pmatrix}$$

which depend on the corresponding distribution functions f_i and f_j.

Remark 3.2.3. *The shape of the target functions is characterized by* (3.31)*. The corresponding value of the target functions comes by evaluation through the parameters* $\boldsymbol{\lambda} = (\lambda^{(0)}, \boldsymbol{\lambda}^{(1)}, \lambda^{(2)})^\top$. *For convenience, we provide conversion formulas for the parameters.*

In the case $\mathbf{w} = \mathbf{v}$ *with velocity-dependent collision frequencies, the parameters* $\boldsymbol{\lambda}$ *coincide exactly with the parameters from Section 2.2.2, and no conversion is needed.*

In the quantum case $\mathbf{w} = \mathbf{p}$, *we presented the model in Section 2.3 with parameters* $a, \mathbf{b}, c$. *An easy calculation gives the conversion formulae*

$$\begin{pmatrix} a \\ \mathbf{b} \\ c \end{pmatrix} = \begin{pmatrix} -\lambda^{(2)} \\ -\frac{\boldsymbol{\lambda}^{(1)}}{2\lambda^{(2)}} \\ -\lambda^{(0)} + \frac{m}{4}\frac{|\boldsymbol{\lambda}^{(1)}|^2}{\lambda^{(2)}} \end{pmatrix} \iff \boldsymbol{\lambda} = \begin{pmatrix} \lambda^{(0)} \\ \boldsymbol{\lambda}^{(1)} \\ \lambda^{(2)} \end{pmatrix} = \begin{pmatrix} -ma|\mathbf{b}|^2 - c \\ 2a\mathbf{b} \\ -a \end{pmatrix}.$$

The target function $\mathcal{A}$ is a Maxwellian distribution $\mathcal{M}[n, \mathbf{u}, T, m]$ in the very special case $\mathbf{w} = \mathbf{p}$ and $\tau = 0$. (This also holds true for $\mathbf{w} = \mathbf{v}$ and constant in $\mathbf{v}$ collision frequencies.) Only under these circumstances, it is possible to formally convert the parameters $\boldsymbol{\lambda} = (\lambda^{(0)}, \boldsymbol{\lambda}^{(1)}, \lambda^{(2)})^\top$ into the macroscopic quantities $n, \mathbf{u}, T$ in an analytic way, i.e.

$$n = \left(-\frac{\pi}{m\lambda^{(2)}}\right)^{3/2} \exp\left(m\lambda^{(0)} - \frac{m|\boldsymbol{\lambda}^{(1)}|^2}{4\lambda^{(2)}}\right),$$
$$\mathbf{u} = -\frac{\boldsymbol{\lambda}^{(1)}}{2\lambda^2},$$
$$T = -\frac{1}{2\lambda^{(2)}},$$

respective

$$\lambda^{(0)} = \frac{1}{m}\log\left(\frac{n}{(2\pi\frac{T}{m})^{3/2}}\right) - \frac{|\mathbf{u}|^2}{2T},$$
$$\boldsymbol{\lambda}^{(1)} = \frac{\mathbf{u}}{T},$$
$$\lambda^{(2)} = -\frac{1}{2T}.$$

The model is required to satisfy the conservation properties for intra-species interactions

$$\int \nu_{ii}(\mathbf{w})(\mathcal{A}_{ii,\tau_i}(\mathbf{w}) - f_i)\,\mathbf{a}_i(\mathbf{w})\,\mathrm{d}\mathbf{w} = 0, \tag{3.32}$$

the conservation properties for inter-species interactions

$$\begin{aligned}
\int \nu_{11}(\mathbf{w})\,\mathrm{a}_1^{(0)}(\mathbf{w})\,(\mathcal{A}_{12,\tau_1} - f_1)\,\mathrm{d}\mathbf{w} = 0, \quad & \int \nu_{22}(\mathbf{w})\,\mathrm{a}_2^{(0)}(\mathbf{w})\,(\mathcal{A}_{21,\tau_2} - f_2)\,\mathrm{d}\mathbf{w} = 0,\\
\int \nu_{12}(\mathbf{w})\,\mathbf{a}_1^{(1)}(\mathbf{w})\,(\mathcal{A}_{12,\tau_1} - f_1)\,\mathrm{d}\mathbf{w} + & \int \nu_{21}(\mathbf{w})\,\mathbf{a}_2^{(1)}(\mathbf{w})\,(\mathcal{A}_{21,\tau_2} - f_2)\,\mathrm{d}\mathbf{w} = 0,\\
\int \nu_{12}(\mathbf{w})\,\mathrm{a}_1^{(2)}(\mathbf{w})\,(\mathcal{A}_{12,\tau_1} - f_1)\,\mathrm{d}\mathbf{w} + & \int \nu_{21}(\mathbf{w})\,\mathrm{a}_2^{(2)}(\mathbf{w})\,(\mathcal{A}_{21,\tau_2} - f_2)\,\mathrm{d}\mathbf{w} = 0
\end{aligned} \tag{3.33}$$

and the correct behavior of the entropy

$$\mathcal{H}[\{f_1, f_2\}](\mathbf{w}) = \int [h_{\tau_1}(f_1(\mathbf{w})) + h_{\tau_2}(f_2(\mathbf{w}))]\,\mathrm{d}\mathbf{w}$$

with

$$h_{\tau_i}(f_i) = f_i \log(f_i) - f_i + \tau_i(1 - \tau_i f_i)\log(1 - \tau_i f_i) - \tau_i(1 - \tau_i f_i). \tag{3.34}$$

The system of equations (3.29) shall be solved numerically. At the discrete level, the conservation properties (3.32)–(3.33) and the correct entropy behavior are required as well. In the following, we present our scheme in details.[5] We start with the discretization of the time variable.

3.3 Time discretization

We pursue IMEX time discretizations where, in (3.29), $\mathcal{T}$ is treated explicitly and $\mathcal{R}_i$ is treated implicitly. Our approach works for multi-species BGK quantum models, respective for multi-species BGK equations equipped with a broad class of collision frequencies. However, the treatment of the collision frequencies relies on Assumption 2.2.11. It states that the collision frequencies depend only implicitly on space and time via a dependence on the mass densities $\rho_i(\mathbf{x}, t)$, the mixture mean velocity

$$\mathbf{u}_{\text{mix}}(\mathbf{x}, t) = \frac{\rho_1 \mathbf{u}_1 + \rho_1 \mathbf{u}_2}{\rho_1 + \rho_2} \tag{3.35}$$

and the mixture temperature

$$\begin{aligned} T_{\text{mix}}(\mathbf{x}, t) &= \frac{1}{n_1 + n_2} \left(n_1 T_1 + n_2 T_2 + \frac{1}{3} \sum_{i=1}^{2} \rho_i (|\mathbf{u}_i|^2 - |\mathbf{u}_{\text{mix}}|^2) \right) \\ &= \frac{1}{n_1 + n_2} \left(n_1 T_1 + n_2 T_2 + \frac{\rho_1 \rho_2}{3(\rho_1 + \rho_2)} (|\mathbf{u}_1 - \mathbf{u}_2|^2) \right), \end{aligned}$$

introduced in Definition 2.1.16. Since the collisional process conserves these quantities, the collision frequencies ν_{ij} are independent of time in the space homogeneous setting. For example, given $t_\ell = \ell \Delta t$ for $\ell \in \mathbb{N}_0$ a simple update of $f_i^\ell \approx f_i(\mathbf{x}, \mathbf{w}, t_\ell)$ from t_ℓ to $t_{\ell+1}$ uses the approximation

$$\mathcal{R}_i(f_i^{\ell+1}, f_j^{\ell+1}) \approx \nu_{ii}^\ell \left(\mathcal{A}_{ii,\tau_i}^{\ell+1} - f_i^{\ell+1} \right) + \nu_{ij}^\ell \left(\mathcal{A}_{ij,\tau_i}^{\ell+1} - f_i^{\ell+1} \right). \tag{3.36}$$

The discrete target functions $\mathcal{A}_{ii,\tau_i}^{\ell+1}$ and $\mathcal{A}_{ij,\tau_i}^{\ell+1}$ are described in detail in Section 3.3.5. Here, we only mention that they depend on $f_i^{\ell+1}, f_j^{\ell+1}, \nu_{ii}^\ell$ and ν_{ij}^ℓ via the solution of a convex minimization problem. The evaluation of the collision frequencies at time step t_ℓ is justified because of Assumption 2.2.11 which gives

$$\nu_{ij}^{\ell+1} = \nu_{ij}(\rho_i^{\ell+1}, \rho_j^{\ell+1}, \mathbf{u}_{\text{mix}}^{\ell+1}, T_{\text{mix}}^{\ell+1}) = \nu_{ij}(\rho_i^\ell, \rho_j^\ell, \mathbf{u}_{\text{mix}}^\ell, T_{\text{mix}}^\ell) = \nu_{ij}^\ell. \tag{3.37}$$

However, in more general settings, lagging the collision frequencies in this way may cause a drop in temporal order for an otherwise high-order scheme [Low04].

In the following sections, we formulate several time discretization methods. The main aspect is provided in Section 3.3.5 where we present our general implicit solver in order to determine $\mathcal{A}_{ii,\tau_i}^{\ell+1}$ and $\mathcal{A}_{ij,\tau_i}^{\ell+1}$. We want to emphasize that the implicit solver is not limited to the presented time discretizations, but it can be used for many more discretizations in a straight-forward way.

[5] In Appendix C, we provide pseudo codes for our scheme.

3.3.1 First-order splitting

We split the system of PDEs (3.29) into a relaxation step and the transport step. The order of the steps is not fixed. However, we expect that the relaxation processes take place on smaller time scales, so we first perform the relaxation step and afterwards the transport step.

Relaxation step The relaxation step is executed in each spatial cell using the Backward Euler method (3.8)

$$\frac{f_i^* - f_i^\ell}{\Delta t} = \mathcal{R}_i(f_i^*, f_j^*), \tag{3.38}$$

which can be rewritten to express f_i^* as the convex combination

$$f_i^* = c_i^\ell f_i^\ell + c_i^\ell \Delta t(\nu_{ii}^\ell \mathcal{A}_{ii,\tau_i}^* + \nu_{ij}^\ell \mathcal{A}_{ij,\tau_i}^*) \tag{3.39}$$

with

$$c_i^\ell = \frac{1}{1 + \Delta t(\nu_{ii}^\ell + \nu_{ij}^\ell)}. \tag{3.40}$$

If $\mathcal{A}_{ii,\tau_i}^*$ and $\mathcal{A}_{ij,\tau_i}^*$ can be expressed as functions of f_i^ℓ and f_j^ℓ, then (3.39) provides an explicit update formula for $f_i^{\ell'}$. In Section 3.3.5, we show how to determine $\mathcal{A}_{ii,\tau_i}^*$ and $\mathcal{A}_{ij,\tau_i}^*$ while preserving the conservation properties (3.32) and (3.33) at the discrete level.

It is also possible to consider the intra-species and inter-species relaxation process independently. However, we prefer to consider the entire relaxation process as the different interactions are already represented by the model.

Transport step We solve the transport in x for $f_i^{\ell+1}$ by the Forward Euler method (3.2) with initial data f_i^*:

$$\frac{f_i^{\ell+1} - f_i^*}{\Delta t} + \mathcal{T}(f_i^*) = 0. \tag{3.41}$$

Details on the numerical approximation of $\mathcal{T}$ are given in Section 3.4.

3.3.2 First-order IMEX Runge-Kutta

We use the combination of a Backward Euler method and a Forward Euler method for a first-order IMEX RK scheme. It is described in [ARS97] where one also finds some stability considerations. Applying the method to (3.29) yields

$$f_i^{\ell+1} = f_i^\ell - \Delta t\, \mathcal{T}(f_i^\ell) + \Delta t\, \mathcal{R}_i(f_i^{\ell+1}, f_j^{\ell+1})$$

which can be rewritten into a convex combination

$$f_i^{\ell+1} = c_i^\ell G_i^\ell + c_i^\ell \Delta t\, \nu_{ii}^\ell \mathcal{A}_{ii,\tau_i}^{\ell+1} + c_i^\ell \Delta t\, \nu_{ij}^\ell \mathcal{A}_{ij,\tau_i}^{\ell+1} \tag{3.42}$$

with

$$G_i^\ell = f_i^\ell - \Delta t\, \mathcal{T}(f_i^\ell)$$

and

$$c_i^\ell = \frac{1}{1 + \Delta t(\nu_{ii}^\ell + \nu_{ij}^\ell)}.$$

The update (3.42) is explicit provided that $\mathcal{A}_{ii,\tau_i}^{\ell+1}$ and $\mathcal{A}_{ij,\tau_i}^{\ell+1}$ can be expressed as functions of f_i^ℓ. This is discussed in Section 3.3.5.

3.3.3 Second-order Strang splitting

We arrange the relaxation steps and the transport steps presented in Section 3.3.1 according to the Strang splitting method, see Definition 3.1.9.

Additionally, each step itself needs to be solved in a second-order manner. The relaxation step is surrounded by transport steps with a half step size following the second-order explicit Heun's method (3.4). This will be described in Section 3.4.

Given f_i^* after the transport step with a half step size. For the relaxation step, we apply the second-order implicit trapezoidal method (3.10), resulting in

$$f_i^{**} = c_i^{**} G_i^* + c_i^{**} \frac{\Delta t}{2} (\nu_{ii}^{**} \mathcal{A}_{ii,\tau_i}^{**} + \nu_{ij}^{**} \mathcal{A}_{ij,\tau_i}^{**}) \tag{3.43}$$

with

$$G_i^* = f_i^* + \frac{\Delta t}{2} (\nu_{ii}^* (\mathcal{A}_{ii,\tau_i}^* - f_i^*) + \nu_{ij}^* (\mathcal{A}_{ij,\tau_i}^* - f_i^*))$$

and

$$c_i^{**} = \frac{1}{1 + \frac{\Delta t}{2}(\nu_{ii}^{**} + \nu_{ij}^{**})}.$$

Importantly, the values of $\mathcal{A}_{ij,\tau_i}^*$ for the given data G_i^* need to be determined accurately in order to guarantee the conservation properties. In general, these values are not known because of the previous transport step.

Given f_i^{**}, another transport step with a half step size gives $f_i^{\ell+1} = f_i^{***}$.

Unfortunately, accuracy properties of higher-order splitting schemes may break down when the collision frequencies become large; see [Jin95] for a case where the nominally second-order Strang splitting reduces to a first-order scheme in the presence of large collision frequencies.

3.3.4 Second-order IMEX Runge-Kutta

For a second-order IMEX RK scheme, we use the following Butcher tableaux (3.13), (3.7),

$$\begin{array}{c|ccc} 0 & & & \\ \gamma & 0 & \gamma & \\ 1 & 0 & 1-\gamma & \gamma \\ \hline & 0 & 1-\gamma & \gamma \end{array} \qquad \begin{array}{c|ccc} 0 & & & \\ \gamma & \gamma & & \\ 1 & \delta & 1-\delta & 0 \\ \hline & \delta & 1-\delta & 0 \end{array} \tag{3.44}$$

with

$$\gamma = 1 - \frac{\sqrt{2}}{2} \quad \text{and} \quad \delta = 1 - \frac{1}{2\gamma}, \tag{3.45}$$

see [ARS97]. This IMEX RK scheme is L-stable and GSA. The left table is used for the relaxation step, and the right table is used for the transport step.

Applying the method to (3.29) results in the following stages and updates:

$$f_i^{[1]} = f_i^\ell - \gamma \Delta t\, \mathcal{T}(f_i^\ell) + \gamma \Delta t\, \mathcal{R}_i(f_i^{[1]}, f_j^{[1]}), \tag{3.46a}$$

$$f_i^{[2]} = f_i^\ell - \delta \Delta t\, \mathcal{T}(f_i^\ell) - (1-\delta)\Delta t\, \mathcal{T}(f_i^{[1]}) + (1-\gamma)\Delta t\, \mathcal{R}_i(f_i^{[1]}, f_j^{[1]}) + \gamma \Delta t \mathcal{R}_i(f_i^{[2]}, f_j^{[2]}), \tag{3.46b}$$

$$f_i^{\ell+1} = f_i^{[2]}. \tag{3.46c}$$

Using

$$c_i^{[r]} = \frac{1}{1 + \gamma \Delta t (\nu_{ii}^{[r]} + \nu_{ij}^{[r]})}, \tag{3.47}$$

we can rewrite (3.46a) and (3.46b) as convex combination of three terms

$$f_i^{[1]} = c_i^{[1]} G_i^{[1]} + c_i^{[1]} \gamma \Delta t\, \nu_{ii}^{[1]} \mathcal{A}_{ii,\mathcal{T}_i}^{[1]} + c_i^{[1]} \gamma \Delta t\, \nu_{ij}^{[1]} \mathcal{A}_{ij,\mathcal{T}_i}^{[1]} \tag{3.48a}$$

$$f_i^{[2]} = c_i^{[2]} G_i^{[2]} + c_i^{[2]} \gamma \Delta t\, \nu_{ii}^{[2]} \mathcal{A}_{ii,\mathcal{T}_i}^{[2]} + c_i^{[2]} \gamma \Delta t\, \nu_{ij}^{[2]} \mathcal{A}_{ij,\mathcal{T}_i}^{[2]}, \tag{3.48b}$$

where the quantities

$$G_i^{[1]} = f_i^\ell - \Delta t\, \gamma\, \mathcal{T}(f_i^\ell) \tag{3.49a}$$

$$G_i^{[2]} = f_i^\ell - \Delta t\, \delta\, \mathcal{T}(f_i^\ell) - \Delta t\,(1-\delta)\mathcal{T}(f_i^{[1]}) + \Delta t\,(1-\gamma)\mathcal{R}_i(f_i^{[1]}, f_j^{[1]}) \tag{3.49b}$$

depend on known data. The collision frequencies $\nu_{ii}^{[r]}$, $\nu_{ij}^{[r]}$ and constants $c_i^{[r]}$ are evaluated at the intermediate steps $G_i^{[r]}$. This option maintains second-order accurate as long as Assumption 2.2.11 applies where we use that collisional processes do not change the collision frequencies.

The main computational challenge in each stage of (3.48) is to determine the parameters of the target functions. In the following section, we explain how to manage this.

3.3.5 General implicit solver

In this section, we describe a method how to deal with the implication of the target functions which is required for any of the previously presented time discretizations. We write the implicit updates in (3.39), (3.42), (3.43) and (3.48) in a generic steady state form

$$f_i = c_i G_i + c_i \gamma \Delta t (\nu_{ii} \mathcal{A}_{ii,\tau_i} + \nu_{ij} \mathcal{A}_{ij,\tau_i}) \tag{3.50}$$

where $\mathcal{A}_{ii,\tau_i}$ and $\mathcal{A}_{ij,\tau_i}$ are the unique target functions associated to f_i,

$$c_i = \frac{1}{1 + \gamma \Delta t (\nu_{ii} + \nu_{ij})}, \tag{3.51}$$

and G_i is a known function. We aim to express $\mathcal{A}_{ii,\tau_i}$ and $\mathcal{A}_{ij,\tau_i}$ as functions of G_i and G_j so that (3.50) provides an *explicit* update formula for f_i even though the target functions depend on f_i via moment equations. In the following, we explain how to manage this. The resulting general implicit solver can be applied to any time discretization method as long as the update can be written in the form of (3.50).

Applying the conservation properties (3.32) and (3.33) to (3.50) gives

$$\begin{aligned}
&\int \nu_{11} \mathcal{A}_{11,\tau_1} \, \mathbf{a}_1(\mathbf{w}) \, \mathrm{d}\mathbf{w} + \int \nu_{22} \mathcal{A}_{22,\tau_2} \, \mathbf{a}_2(\mathbf{w}) \, \mathrm{d}\mathbf{w} \\
&\qquad + \int \nu_{12} \mathcal{A}_{12,\tau_1} \, \mathbf{a}_1(\mathbf{w}) \, \mathrm{d}\mathbf{w} + \int \nu_{21} \mathcal{A}_{21,\tau_2} \, \mathbf{a}_2(\mathbf{w}) \, \mathrm{d}\mathbf{w} \\
&\overset{(3.32),(3.33)}{=} \int \nu_{11} \psi_1 \, \mathbf{a}_1(\mathbf{w}) \, \mathrm{d}\mathbf{w} + \int \nu_{22} \psi_2 \, \mathbf{a}_2(\mathbf{w}) \, \mathrm{d}\mathbf{w} + \\
&\qquad + \int \nu_{12} \psi_1 \, \mathbf{a}_1(\mathbf{w}) \, \mathrm{d}\mathbf{w} + \int \nu_{21} \psi_2 \, \mathbf{a}_2(\mathbf{w}) \, \mathrm{d}\mathbf{w} \\
&\overset{(3.50)}{=} \int \nu_{11} c_1 \left[G_1 + \Delta t \, \gamma \nu_{11} A_{11,\tau_1} + \Delta t \, \gamma \nu_{12} \mathcal{A}_{12,\tau_1} \right] \mathbf{a}_1(\mathbf{w}) \, \mathrm{d}\mathbf{w} \\
&\qquad + \int \nu_{22} c_2 \left[G_2 + \Delta t \, \gamma \nu_{22} A_{22,\tau_2} + \Delta t \, \gamma \nu_{21} \mathcal{A}_{21,\tau_2} \right] \mathbf{a}_2(\mathbf{w}) \, \mathrm{d}\mathbf{w} \\
&\qquad + \int \nu_{12} c_1 \left[G_1 + \Delta t \, \gamma \nu_{11} A_{11,\tau_1} + \Delta t \, \gamma \nu_{12} \mathcal{A}_{12,\tau_1} \right] \mathbf{a}_1(\mathbf{w}) \, \mathrm{d}\mathbf{w} \\
&\qquad + \int \nu_{21} c_2 \left[G_2 + \Delta t \, \gamma \nu_{22} A_{22,\tau_2} + \Delta t \, \gamma \nu_{21} \mathcal{A}_{21,\tau_2} \right] \mathbf{a}_2(\mathbf{w}) \, \mathrm{d}\mathbf{w}.
\end{aligned} \tag{3.52}$$

After sorting terms, we arrive at the following moment equations

$$\begin{aligned}
&\int c_1 \left(\nu_{11} \mathcal{A}_{11,\tau_1} + \nu_{12} \mathcal{A}_{12,\tau_1} \right) \mathbf{a}_1(\mathbf{w}) \, \mathrm{d}\mathbf{w} + \int c_2 \left(\nu_{21} \mathcal{A}_{21,\tau_2} + \nu_{22} \mathcal{A}_{22,\tau_2} \right) \mathbf{a}_2(\mathbf{w}) \, \mathrm{d}\mathbf{w} \\
&\qquad = \int c_1 \left(\nu_{11} + \nu_{12} \right) G_1 \mathbf{a}_1(\mathbf{w}) + \int c_2 \left(\nu_{22} + \nu_{21} \right) G_2 \mathbf{a}_2(\mathbf{w}) \, \mathrm{d}\mathbf{w},
\end{aligned} \tag{3.53}$$

which provide a set of constraints to determine $\mathcal{A}_{11,\tau_1}$, $\mathcal{A}_{12,\tau_1}$, $\mathcal{A}_{21,\tau_2}$ and $\mathcal{A}_{22,\tau_2}$ from the given data G_1 and G_2.

The system of equations (3.53) can be solved in an elegant way by realizing that these constraints in (3.53) represent first-order optimality conditions associated to the minimization of the convex function

$$\varphi_{\text{tot}}(\boldsymbol{\alpha}_1, \boldsymbol{\alpha}_2, \boldsymbol{\alpha}) = \mathfrak{H}[\mathcal{A}_{11,\tau_1}, \mathcal{A}_{12,\tau_1}, \mathcal{A}_{21,\tau_2}, \mathcal{A}_{22,\tau_2}](\mathbf{w}) + \boldsymbol{\mu}_{11} \cdot \boldsymbol{\alpha}_1 + \boldsymbol{\mu}_{22} \cdot \boldsymbol{\alpha}_2 + \boldsymbol{\mu} \cdot \boldsymbol{\alpha} \tag{3.54}$$

with

$$\mathfrak{H}[\mathcal{A}_{11,\tau_1}, \mathcal{A}_{12,\tau_1}, \mathcal{A}_{21,\tau_2}, \mathcal{A}_{22,\tau_2}](\mathbf{w}) = \sum_{i,j} \int c_i \nu_{ij} \mathfrak{h}_{\tau_i}[\mathcal{A}_{ij,\tau_i}] \, \mathrm{d}\mathbf{w}$$

where

$$\mathfrak{h}_{\tau_i}[\mathcal{A}_{ij,\tau_i}] = \frac{\log(1 - \tau_i \mathcal{A}_{ij,\tau_i})}{\tau_i} = \begin{cases} -\mathcal{A}_{ij,0} & \text{for } \tau_i = 0, \\ \log(1 - \mathcal{A}_{ij,1}) & \text{for } \tau_i = +1, \\ -\log(1 + \mathcal{A}_{ij,-1}) & \text{for } \tau_i = -1. \end{cases} \tag{3.55}$$

Moreover, $\boldsymbol{\alpha}_i = (\alpha_i^{(0)}, \boldsymbol{\alpha}_i^{(1)}, \alpha_i^{(2)})^\top$;

$$\boldsymbol{\mu}_{ii} = \begin{pmatrix} \mu_{ii}^{(0)} \\ \boldsymbol{\mu}_{ii}^{(1)} \\ \mu_{ii}^{(2)} \end{pmatrix} = \int c_i \nu_{ii} G_i \mathbf{a}_i(\mathbf{w}) \, \mathrm{d}\mathbf{w} \tag{3.56}$$

for $i = 1, 2$; $\boldsymbol{\alpha} = (\alpha_{12}^{(0)}, \alpha_{21}^{(0)}, \boldsymbol{\alpha}^{(1)}, \alpha^{(2)})^\top$; and

$$\boldsymbol{\mu} = \begin{pmatrix} \mu_{12}^{(0)} \\ \mu_{21}^{(0)} \\ \boldsymbol{\mu}^{(1)} \\ \mu^{(2)} \end{pmatrix} = \int \left[\begin{pmatrix} \mathrm{a}_1^{(0)}(\mathbf{w}) \\ 0 \\ \mathbf{a}_1^{(1)}(\mathbf{w}) \\ \mathrm{a}_1^{(2)}(\mathbf{w}) \end{pmatrix} c_1 \nu_{12} G_1 + \begin{pmatrix} 0 \\ \mathrm{a}_2^{(0)}(\mathbf{w}) \\ \mathbf{a}_2^{(1)}(\mathbf{w}) \\ \mathrm{a}_2^{(2)}(\mathbf{w}) \end{pmatrix} c_2 \nu_{21} G_2 \right] \mathrm{d}\mathbf{w}. \tag{3.57}$$

The minimization problem can be decoupled as follows which makes the procedure better handable.

Proposition 3.3.1. *The components of the minimizer of (3.54) can be found by minimizing the following three convex functions independently:*

$$\varphi_i(\boldsymbol{\alpha}_i) = \int c_i \nu_{ii} \mathfrak{h}_{\tau_i}[\mathcal{A}_{ii,\tau_i}] \, \mathrm{d}\mathbf{w} + \boldsymbol{\mu}_{ii} \cdot \boldsymbol{\alpha}_i \quad \textit{for} \quad i = 1, 2 \quad \textit{and} \tag{3.58}$$

$$\varphi(\boldsymbol{\alpha}) = \int (c_1 \nu_{12} \mathfrak{h}_{\tau_1}[\mathcal{A}_{12,\tau_1}] + c_2 \nu_{21} \mathfrak{h}_{\tau_2}[\mathcal{A}_{21,\tau_2}]) \, \mathrm{d}\mathbf{w} + \boldsymbol{\mu} \cdot \boldsymbol{\alpha} \tag{3.59}$$

and the minimum of (3.54) *is the sum of their minima.*

Proof. The statement is trivial because (3.54) can be written as the sum of the three potential functions, whose arguments are independent; that is, $\varphi_{\text{tot}}(\boldsymbol{\alpha}_1, \boldsymbol{\alpha}_2, \boldsymbol{\alpha}) = \varphi_1(\boldsymbol{\alpha}_1) + \varphi_2(\boldsymbol{\alpha}_2) + \varphi_1(\boldsymbol{\alpha})$. $\square$

The minimum of each potential function in (3.58) and (3.59) is found using Newton's method for convex optimization. More details are given in Section 3.5.

The above procedure is appealing because the minimization problems in (3.58) and (3.59) are numerical analogs of the dual problems in the analytic case. We emphasize this fact for both models in Sections 2.2 and 2.3 and discuss the existence and uniqueness of solutions to (3.58) and (3.59).

For $\mathbf{w} = \mathbf{v}$ (the model with velocity-dependent collision frequencies in Section 2.2), the potential functions (3.58), respective (3.59) read

$$\varphi_i(\boldsymbol{\alpha}_i) = -\int c_i \nu_{ii} \mathcal{G}_{ii} \, \mathrm{d}\mathbf{v} + \boldsymbol{\mu}_{ii} \cdot \boldsymbol{\alpha}_i, \tag{3.60}$$

$$\varphi(\boldsymbol{\alpha}) = -\int (c_1 \nu_{12} \mathcal{G}_{12} + c_2 \nu_{21} \mathcal{G}_{21}) \, \mathrm{d}\mathbf{v} + \boldsymbol{\mu} \cdot \boldsymbol{\alpha}, \tag{3.61}$$

where $\boldsymbol{\alpha}_i = (\alpha_i^{(0)}, \boldsymbol{\alpha}_i^{(1)}, \alpha_i^{(2)})^\top \in \mathbb{R} \times \mathbb{R}^3 \times \mathbb{R}^-$ and $\boldsymbol{\alpha} = (\alpha_{12}^{(0)}, \alpha_{21}^{(0)}, \boldsymbol{\alpha}^{(1)}, \alpha^{(2)})^\top \in \mathbb{R} \times \mathbb{R} \times \mathbb{R}^3 \times \mathbb{R}^-$. The dual problems in the analytic case are given in (2.77) and (2.81), respectively. Indeed, the temporal discretization simply introduces the additional weights $c_i^\ell \to 1$ as $\Delta t \to 0$. Hence, the existence and uniqueness are guaranteed by the theory in [HHK+21]. Essentially, one needs only replace the collision frequencies ν_{ij} by

$$\nu_{ij}^* = c_i \nu_{ij} = \frac{\nu_{ij}}{1 + \gamma \Delta t (\nu_{ii} + \nu_{ij})} \tag{3.62}$$

and then verify that ν_{ij}^* satisfies the conditions used in [HHK+21]. These conditions are mild integrability conditions. Whenever they are satisfied by ν_{ij}, they are easily satisfied by ν_{ij}^*, as well, because $0 < c_i < 1$.

For $\mathbf{w} = \mathbf{p}$ (the quantum model in Section 2.3), it is

$$\varphi_i(\boldsymbol{\alpha}_i) = \int c_i \mathfrak{h}_{\tau_i}[\mathcal{E}_{ii,\tau_i}] \, \mathrm{d}\mathbf{p} + \boldsymbol{\mu}_{ii} \cdot \boldsymbol{\alpha}_i, \tag{3.63}$$

$$\varphi(\boldsymbol{\alpha}) = \int (c_1 \mathfrak{h}_{\tau_1}[\mathcal{E}_{12,\tau_1}] + c_2 \mathfrak{h}_{\tau_2}[\mathcal{E}_{21,\tau_2}]) \, \mathrm{d}\mathbf{p} + \boldsymbol{\mu} \cdot \boldsymbol{\alpha}. \tag{3.64}$$

The consistency of the quantum model in Section 2.3 is proven by algebraic considerations. However, we realize that we can reformulate the modelling problem by using Lagrange functionals.[6] For intra-species interactions, the Lagrange functional reads

$$L_{ii}(g, \boldsymbol{\alpha}) = \int h_{\tau_i}(g) \, \mathrm{d}\mathbf{p} - \boldsymbol{\alpha} \cdot \int \mathbf{a}_i(\mathbf{p})(g - f_i) \, \mathrm{d}\mathbf{p}$$

6 The first-order optimality condition, for instance, gives us the shape (3.31) of the target functions.

with h_{τ_i} defined in (3.34). The multipliers solve the corresponding dual problem

$$\boldsymbol{\lambda}_{ii} = \underset{\boldsymbol{\alpha}\in\Lambda_{ii}}{\operatorname{argmin}} \int \mathfrak{h}_{\tau_i}[\mathcal{E}_{ii,\tau_i}(\boldsymbol{\alpha})]\,\mathrm{d}\mathbf{p} + \boldsymbol{\alpha}\cdot\int \mathbf{a}_i(\mathbf{p}) f_i\,\mathrm{d}\mathbf{p} \tag{3.65}$$

where $\Lambda_{ii} = \{\boldsymbol{\lambda}_i \in \mathbb{R}^5 \mid \int \mathcal{A}_{ii,\tau_i}(\boldsymbol{\lambda}_i)(1+|\mathbf{p}|^2)\,\mathrm{d}\mathbf{p} < \infty\}$. The dual problem for inter-species interaction can be obtained analogously:

$$\begin{aligned}(\boldsymbol{\lambda}_{12},\boldsymbol{\lambda}_{21}) = \underset{(\boldsymbol{\alpha}_{12},\boldsymbol{\alpha}_{21})\in\Lambda_{12}}{\operatorname{argmin}} \Bigg\{ &\int (\mathfrak{h}_{\tau_i}[\mathcal{E}_{12,\tau_1}] + \mathfrak{h}_{\tau_2}[\mathcal{E}_{21,\tau_2}])\,\mathrm{d}\mathbf{p} \\ &+ \alpha_{12}^{(0)} \int f_1\,\mathrm{d}\mathbf{v} + \alpha_{21}^{(0)} \int f_2\,\mathrm{d}\mathbf{p} \\ &+ \boldsymbol{\alpha}^{(1)}\cdot \int \mathbf{p}(f_1+f_2)\,\mathrm{d}\mathbf{p} \\ &+ \alpha^{(2)} \int |\mathbf{p}|^2 \left(\frac{1}{2m_1} f_1 + \frac{1}{2m_2} f_2\right)\mathrm{d}\mathbf{p}\Bigg\}\end{aligned} \tag{3.66}$$

where $\Lambda_{12} = \{(\boldsymbol{\lambda}_{12},\boldsymbol{\lambda}_{21}) \mid \int \mathcal{A}_{ij,\tau_i}(\boldsymbol{\lambda}_{ij})(1+|\mathbf{p}|^2)\,\mathrm{d}\mathbf{p} < \infty \text{ for } i,j = 1,2; i\neq j\}$. We see the close relationship of (3.63) with (3.65) and of (3.64) with (3.66), respectively. The theory in Section 2.3 guarantees a unique solution to these problems in the analytic case. Since c_i is independent of $\mathbf{p}$ (we assume constant collision frequencies $\nu_{ij} = 1$), it does not affect the minimization of (3.63) and (3.64), respectively, such that there exists a unique solution in the discrete case as well.

Remark 3.3.2 (Determine $\mathcal{A}^{\ell}_{ij,\tau_i}$). *The main task in our scheme is to determine the implication of the target functions, i.e. to determine $\mathcal{A}^{\ell+1}_{ij,\tau_i}$. However, especially for the Strang splitting, one also needs to determine the current value of the target functions accurately. This can be done by solving the corresponding dual problems themselves numerically.*

To conclude, we summarize the main ideas of this section. The update of the distribution functions needs to be written in the form of the convex combination in (3.50). This is possible for many time discretization techniques even if we only presented four different schemes. The essential step is to determine the implictly evaluated target functions which turns (3.50) into an explicit update formula. We achieve this by a minimization procedure which mimics the entropy behavior and conservation properties at the discrete level. The well-posedness of the minimization problem is guaranteed by the theoretical properties of the corresponding model.

In the following section, we discuss the discretization of the space variable.

3.4 Space discretization

As already mentioned, we assume a slab geometry and reduce the physical space dimension to one dimension; we set $x = x^{(1)}$. We divide the spatial domain $[x_{\min}, x_{\max}]$ into uniform cells $I_k = [x_k - \frac{\Delta x}{2}, x_k + \frac{\Delta x}{2}]$ for $k \in \{0,\dots,K\}$.

We employ a FV framework that tracks approximate cell-averaged quantities

$$f_{i,k}^{\ell} \approx \frac{1}{\Delta x} \int_{I_k} f_i(x, \mathbf{w}, t^{\ell})\, \mathrm{d}x. \tag{3.67}$$

To second order, we evaluate the spatial cell centers and do not need a further reconstruction. To approximate the relaxation operator, we use the second-order approximation

$$\mathcal{R}_{i,k}^{\ell} = \mathcal{R}_i(f_{i,k}^{\ell}, f_{j,k}^{\ell}) \approx \frac{1}{\Delta x} \int_{I_k} \mathcal{R}\left(f_i(x, \mathbf{w}, t^{\ell}), f_j(x, \mathbf{w}, t^{\ell}) \right) \mathrm{d}x. \tag{3.68}$$

This becomes more delicate and needs to be treated with care for higher-order approximations.

3.4.1 Numerical fluxes

The transport operator $\mathcal{T}$ is discretized with numerical fluxes $\mathscr{F}_{k+\frac{1}{2}}$ by

$$\mathcal{T}(g) \approx \mathcal{T}_k(g) = \frac{1}{\Delta x}\left(\mathscr{F}_{k+\frac{1}{2}}(g) - \mathscr{F}_{k-\frac{1}{2}}(g) \right) \tag{3.69}$$

for any grid function $g = \{g_k\}$. There exist many different approaches for $\mathscr{F}_{k+\frac{1}{2}}$. We use numerical fluxes which are already established in the literature for kinetic equations [MS04], i.e.

$$\mathscr{F}_{k+\frac{1}{2}}(g) = \frac{v^{(1)}}{2}(g_{k+1} + g_k) - \frac{|v^{(1)}|}{2}\left(g_{k+1} - g_k - \phi_{k+\frac{1}{2}}(g) \right) \tag{3.70}$$

where $\phi_{k+\frac{1}{2}}$ is a flux limiter. The choice $\phi_{k+\frac{1}{2}} = 0$ leads to a first-order approximation (the well-known upwind fluxes). A second-order method is provided by

$$\phi_{k+\frac{1}{2}}(g) = \operatorname{minmod}\left((g_k - g_{k-1}), (g_{k+1} - g_k), (g_{k+2} - g_{k+1})\right) \tag{3.71}$$

where

$$\operatorname{minmod}(a, b, c) = \begin{cases} s \min(|a|, |b|, |c|), & \operatorname{sign}(a) = \operatorname{sign}(b) = \operatorname{sign}(c) =: s, \\ 0, & \text{otherwise.} \end{cases} \tag{3.72}$$

For a simple Forward Euler update (3.2) of (3.41), i.e.

$$f_{i,k}^{\ell+1} = f_{i,k}^{\ell} - \frac{\Delta t}{\Delta x}\left(\mathscr{F}_{k+\frac{1}{2}}(f_i^{\ell}) - \mathscr{F}_{k-\frac{1}{2}}(f_i^{\ell}) \right), \tag{3.73}$$

the positivity of f_i is guaranteed by enforcing the CFL condition

$$\Delta t < \alpha \frac{\Delta x}{\max |v^{(1)}|} \tag{3.74}$$

with $\alpha = 1$ for the first-order flux and $\alpha = \frac{2}{3}$ for the second-order flux. (See Proposition 4.2.1.) Applying instead Heun's method (3.4) yields the update formulae

$$
\begin{aligned}
f_{i,k}^{[1]} &= f_{i,k}^{\ell} - \frac{\Delta t}{\Delta x}\left(\mathscr{F}_{k+\frac{1}{2}}(f_i^{\ell}) - \mathscr{F}_{k-\frac{1}{2}}(f_i^{\ell})\right), \\
f_{i,k}^{\ell+1} &= f_{i,k}^{\ell} - \frac{\Delta t}{2\Delta x}\left(\mathscr{F}_{k+\frac{1}{2}}(f_i^{\ell}) - \mathscr{F}_{k-\frac{1}{2}}(f_i^{\ell})\right) - \frac{\Delta t}{2\Delta x}\left(\mathscr{F}_{k+\frac{1}{2}}(f_i^{[1]}) - \mathscr{F}_{k-\frac{1}{2}}(f_i^{[1]})\right).
\end{aligned}
$$

We have introduced the techniques for the discretization in time and space in our scheme. We finally discretize the velocity and momentum variable, respectively, in order to obtain a fully-discrete scheme.

3.5 Velocity discretization

We firstly introduce the construction of an adequate velocity grid.

3.5.1 Velocity grid

We center the discrete velocities $\mathbf{v_q} = (v_{q_1}^{(1)}, v_{q_2}^{(2)}, v_{q_3}^{(3)})^\top$ with $\mathbf{q} = (q_1, q_2, q_3) \in \mathbb{N}_0^3$ around the mixture mean velocity $\mathbf{u}_{\mathrm{mix}}$. For space homogeneous equations, the mixture mean velocity is given by (3.35); for space inhomogeneous problems, we take an a priori guess. We restrict the velocities to a finite cube; that is for each component $p \in \{1, 2, 3\}$

$$
v^{(p)} \in [u_{\mathrm{mix}}^{(p)} - 6v_{\mathrm{th},i}, u_{\mathrm{mix}}^{(p)} + 6v_{\mathrm{th},i}] \tag{3.75}
$$

where $v_{\mathrm{th},i} = \sqrt{\frac{T_{\mathrm{mix}}}{m_i}}$ is the thermal velocity of species i. To ensure an adequate resolution of the velocity domain, the velocity mesh size is chosen to be $\Delta v_i = 0.25\, v_{\mathrm{th},i}$ in each direction. Any quantity is approximated at the nodes by e.g. $f_{\mathbf{q}} \approx f(\mathbf{v_q})$.

For the quantum model, we perform the same discretization by additionally using $\mathbf{p} = m_i \cdot \mathbf{v}$.

We emphasize again the advantage of the multi-species BGK model that it is possible to use different grids for each species/equation [HHM17b]. This feature is a substantial benefit when the species masses, and hence the reference thermal speeds for each species, differ significantly. And this feature relies on the fact that the particles only interact with each other through moments.

3.5.2 Quadrature and discrete moments

All integrals with respect to $\mathbf{w}$ are replaced by discrete sums using the trapezoidal rule, which is known to perform well for smooth, compactly supported functions because they can be viewed as periodic. (See, e.g, [Atk89, Section 5.4, Corollary 1].) Thus

$$\diamondsuit = \int (\cdot)\, \mathrm{d}\mathbf{w} \approx \sum_{\mathbf{q}} \omega_{\mathbf{q}} (\cdot)_{\mathbf{q}} (\Delta w_i)^3 =: \diamondsuit_W, \tag{3.76}$$

where $\omega_{\mathbf{q}} = \omega_{q_1}\omega_{q_2}\omega_{q_3}$ are the weights and

$$\omega_{q_p} = \begin{cases} 1 & \text{if } \min(q_p) < q_p < \max(q_p), \\ \frac{1}{2} & \text{else.} \end{cases} \tag{3.77}$$

Be aware that the quadrature is executed on the corresponding velocity grid and momentum grid, respectively, depending on species i.

The quadrature approximation forces us to distinguish between discrete and continuous moments, especially when determining the discrete local equilibria $\mathcal{A}_{ii,\tau_i,\mathbf{q}}$ and $\mathcal{A}_{ij,\tau_i,\mathbf{q}}$. In fact, the minimization of (3.58) and (3.59) is solved using a discrete-velocity grid and discrete moments $\boldsymbol{\mu}_{i,W}, \boldsymbol{\mu}_W$ as input. Thus, the parameters $\boldsymbol{\lambda}_{ii}$ and $\boldsymbol{\lambda}_{ij}$ are determined such that $\mathcal{A}_{ii,\tau_i,\mathbf{q}}$ and $\mathcal{A}_{ij,\tau_i,\mathbf{q}}$ have the desired discrete moments, and the conservation and entropy properties are fulfilled at the discrete level.

In Section 3.1.3, we summarized a similar approach for the standard, singles-species BGK model, introduced in [Mie00].

We finally give the details for the minimization procedure which completes our scheme.

3.5.3 Optimization algorithm

The minimization of (3.58) and (3.59) is solved by Newton's method with a back-tracking line search [DS96, p. 325], using the `SNESNEWTONLS` solver from PETSc [BAA+21b, BGMS97, BAA+21a]. Newton's method requires the evaluation of gradients and Hessians. Using (3.55), we obtain

$$\nabla_{\boldsymbol{\alpha}_i} \mathfrak{h}_{\tau_i}[\mathcal{A}_{ij,\tau_i}] = -\mathbf{a}_i \mathcal{A}_{ij,\tau_i}$$

and

$$\nabla^2_{\boldsymbol{\alpha}_i} \mathfrak{h}_{\tau_i}[\mathcal{A}_{ij,\tau_i}] = \begin{cases} -\mathcal{A}_{ij,\tau_i}\, \mathbf{a}_i \otimes \mathbf{a}_i & \text{for } \tau_i = 0, \\ -(\mathcal{A}_{ij,\tau_i})^2 \exp(-\boldsymbol{\alpha}_{ij} \cdot \mathbf{a}_i)\, \mathbf{a}_i \otimes \mathbf{a}_i & \text{for } \tau_i = \pm 1. \end{cases}$$

We define the function[7] ζ straight-forwardly by $\nabla^2_{\boldsymbol{\alpha}_i} \mathfrak{h}_{\tau_i}[\mathcal{A}_{ij,\tau_i}] := \zeta(\mathcal{A}_{ij,\tau_i})\, \mathbf{a}_i \otimes \mathbf{a}_i$. This leads to the following gradients:

$$\nabla_{\boldsymbol{\alpha}_i} \varphi_i \approx -(c_i \nu_{ii}\, \mathcal{A}_{ii,\tau_i}\, \mathbf{a}_i)_W + \boldsymbol{\mu}_{i,W}, \tag{3.78}$$

$$\nabla_{\boldsymbol{\alpha}} \varphi \approx -\,(c_1 \nu_{12}\, \mathcal{A}_{12,\tau_1}\, \mathbf{a}_{12})_W - (c_2 \nu_{21}\, \mathcal{A}_{21,\tau_2}\, \mathbf{a}_{21})_W + \boldsymbol{\mu}_W, \tag{3.79}$$

[7] For numerical stability, it is better to implement $\zeta(\mathcal{A}_{ij,\pm 1}) = (\exp(-\boldsymbol{\alpha}_{ij} \cdot \mathbf{a}_i) + \exp(\boldsymbol{\alpha}_{ij} \cdot \mathbf{a}_i) \pm 2)^{-1}$.

and Hessians:

$$\nabla^2_{\boldsymbol{\alpha}_i} \varphi_i \approx (c_i \nu_{ii}\, \zeta(\mathcal{A}_{ii,\tau_i})\, \mathbf{a}_i \otimes \mathbf{a}_i)_W \tag{3.80}$$

$$\nabla^2_{\boldsymbol{\alpha}} \varphi \approx (c_1 \nu_{12}\, \zeta(\mathcal{A}_{12,\tau_1})\, \mathbf{a}_{12} \otimes \mathbf{a}_{12})_W + (c_2 \nu_{21}\, \zeta(\mathcal{A}_{21,\tau_2})\, \mathbf{a}_{21} \otimes \mathbf{a}_{21})_W \tag{3.81}$$

where $\mathbf{a}_{12} = (\mathrm{a}_1^{(0)}, 0, \mathbf{a}_1^{(1)}, \mathrm{a}_1^{(2)})^\top$, $\mathbf{a}_{21} = (0, \mathbf{a}_2)^\top$. For illustration purposes, we give

$$(c_i \nu_{ii}\, \mathcal{A}_{ii,\tau_i}\, \mathbf{a}_i)_W = \sum_{\mathbf{q}} \omega_{\mathbf{q}} (c_i \nu_{ii})_{\mathbf{q}}\, \mathcal{A}_{ii,\tau_i,\mathbf{q}}\, \mathbf{a}_{i,\mathbf{q}} (\Delta w_i)^3,$$

and the input data in (3.56) is computed in a straight-forward way:

$$\boldsymbol{\mu}_{i,W} \approx \sum_{\mathbf{q}} \omega_{\mathbf{q}} (c_i \nu_{ii})_{\mathbf{q}}\, G_{i,\mathbf{q}}\, \mathbf{a}_{i,\mathbf{q}} (\Delta w_i)^3. \tag{3.82}$$

Analogously for the other terms and the input data $\boldsymbol{\mu}_W$ in (3.57).

The Newton method is declared to have converged if one of the standard termination criteria[8] is less than 10^{-14}.

8 For solving $F(x) = 0$, standard termination criteria are: i) $||F|| < \epsilon$, ii) $||F|| < \epsilon ||F(x_C)||$, and iii) $||\Delta x|| < \epsilon ||x||$ for the tolerance ϵ.

Chapter 4

Properties of the numerical scheme

The discretization of an object inevitably introduces errors. Using (prohibitively) small step sizes reduces these errors. Accordingly, one is interested in discretizations which make practicable step sizes possible. The corresponding properties at the discrete level need to be examined analytically before numerical tests can be started.

In this chapter, we validate properties of our scheme by analytical calculations. To start with, we mention the order and stability of our scheme in Section 4.1. Then we review the positivity, conservation properties, and the entropy behavior of our scheme in Sections 4.2–4.5. These statements and proofs can also be found in [HHK+22, BPW22].

4.1 Order and stability

All variables are discretized by first-order or second-order techniques.

When a second-order discretization is needed and a stiff RHS is encountered, we recommend to use the second-order IMEX RK from Section 3.3.4 approach because the Strang splitting in Section 3.3.3 may suffer from an order reduction. [Jin95]

Stability is guaranteed if the corresponding CFL condition (3.74) for the transport steps is statisfied. The relaxation steps are A-stable (or even L-stable). Hence, these do not restrict the choice of the size of the time step regarding stability.

Time step restrictions regarding the positivity of the distribution functions are treated in the following section.

4.2 Positivity of distribution functions

A distribution function is positive by definition, see Definition 2.1.12. We require that a positive initial distribution function is preserved positive by our scheme which can be guaranteed if certain time step restrictions are satisfied.

The first-order time-stepping schemes in Section 3.3.1 and 3.3.2 preserve positivity for both first- and second-order numerical fluxes in space; see Propositions 4.2.1 and 4.2.2. Additionally, we discuss the positivity for the second-order scheme from Section 3.3.3 in Proposition 4.2.3, and give a sufficient criterion. We provide a similar result for the second-order time-stepping scheme from Section 3.3.4 in Proposition 4.2.4 for the space homogeneous case.

Distribution functions for fermions are additionally bounded from above, see Definition 2.3.2. The time discretization methods from Sections 3.3.1–3.3.3 combined with the space discretization from Section 3.4 maintain this property at the discrete level. This is shown in Proposition 4.2.5.

Proposition 4.2.1 (Positivity of first-order splitting scheme). *The first-order time discretization in Section 3.3.1 together with the space discretization described in Section 3.4 is positivity-preserving, provided that*

$$\Delta t \leq \alpha \frac{\Delta x}{\max |v^{(1)}|}, \tag{4.1}$$

with $\alpha = 1$ and $\alpha = \frac{2}{3}$ for the first-order and second-order fluxes, respectively.

Proof. Let $f_{i,k}^{\ell} \geq 0$. For the relaxation step, it holds

$$f_{i,k}^{*} \overset{(3.39)}{=} c_{i,k}^{\ell} f_{i,k}^{\ell} + c_{i,k}^{\ell} \Delta t (\nu_{ii,k}^{\ell} \mathcal{A}_{ii,\tau_i,k}^{*} + \nu_{ij,k}^{\ell} \mathcal{A}_{ij,\tau_i,k}^{*}) \geq 0 \tag{4.2}$$

because $c_{i,k}^{\ell}, \nu_{ii,k}^{\ell}, \nu_{ij,k}^{\ell}, \mathcal{A}_{ii,\tau_i,k}^{*}, \mathcal{A}_{ij,\tau_i,k}^{*} \geq 0$. For the transport step (3.73), we obtain with the first-order fluxes

$$f_{i,k}^{\ell+1} = \left(1 - \frac{\Delta t}{\Delta x} |v^{(1)}|\right) f_{i,k}^{*} + \frac{\Delta t}{\Delta x} |v^{(1)}| f_{i,k-\text{sign}(v^1)}^{*} \geq 0, \tag{4.3}$$

where the last inequality holds in each cell provided that the given CFL condition in (4.1) holds with $\alpha = 1$.

For the second-order fluxes, define $\sigma := \text{sign}(f_{i,k}^{*} - f_{i,k-1}^{*})$. Then one can show that

$$\phi_{k+\frac{1}{2}}(f_i^{*}) \geq \begin{cases} 0 & \text{if} \quad \sigma = +1 \\ f_{i,k+1}^{*} - f_{i,k}^{*} & \text{if} \quad \sigma = -1 \end{cases}, \tag{4.4}$$

$$-\phi_{k-\frac{1}{2}}(f_i^{*}) \geq \begin{cases} f_{i,k-1}^{*} - f_{i,k}^{*} & \text{if} \quad \sigma = +1 \\ 0 & \text{if} \quad \sigma = -1 \end{cases}. \tag{4.5}$$

Hence,

$$
\begin{aligned}
f_{i,k}^{\ell+1} &\overset{(3.73)}{=} \left(1 - \frac{\Delta t}{\Delta x}|v^{(1)}|\right) f_{i,k}^* + \frac{\Delta t}{\Delta x}|v^{(1)}| f_{i,k-\mathrm{sign}(v^{(1)})}^* \\
&\qquad\qquad + \frac{\Delta t}{\Delta x}\frac{|v^{(1)}|}{2}(\phi_{k+\frac{1}{2}}(f_i^*) - \phi_{k-\frac{1}{2}}(f_i^*)) \\
&\geq \left(1 - \frac{\Delta t}{\Delta x}|v^{(1)}|\right) f_{i,k}^* + \frac{\Delta t}{\Delta x}|v^{(1)}| f_{i,k-\mathrm{sign}(v^{(1)})}^* \\
&\qquad\qquad + \frac{\Delta t}{\Delta x}\frac{|v^{(1)}|}{2}\begin{cases}(f_{i,k-1}^* - f_{i,k}^*) & \text{if} \quad \sigma = +1 \\ (f_{i,k+1}^* - f_{i,k}^*) & \text{if} \quad \sigma = -1\end{cases} \\
&= \left(1 - \frac{3}{2}\frac{\Delta t}{\Delta x}|v^{(1)}|\right) f_{i,k}^* + \frac{\Delta t}{\Delta x}|v^{(1)}| f_{i,k-\mathrm{sign}(v^{(1)})}^* \\
&\qquad\qquad + \frac{\Delta t}{\Delta x}\frac{|v^{(1)}|}{2}\begin{cases}f_{i,k-1}^* & \text{if} \quad \sigma = +1 \\ f_{i,k+1}^* & \text{if} \quad \sigma = -1\end{cases} \\
&\geq 0,
\end{aligned}
\tag{4.6}
$$

provided that the CFL condition in (4.1) holds with $\alpha = \frac{2}{3}$. □

This results helps us for the following proposition.

Proposition 4.2.2 (Positivity of first-order IMEX RK scheme). *The first-order time discretization in Section 3.3.2 together with the space discretization described in Section 3.4 is positivity-preserving, provided that*

$$\Delta t \leq \alpha \frac{\Delta x}{\max |v^{(1)}|},$$

with $\alpha = 1$ *and* $\alpha = \frac{2}{3}$ *for the first-order and second-order fluxes, respectively.*

Proof. The statement follows directly from Proposition 4.2.1 and the update formula (3.42). □

It is more difficult to guarantee positivity with a second-order time-stepping. As discussed in Section 3.1.1, unconditionally SSP implicit Runge-Kutta schemes are at most first-order accurate [GST01]. Modified IMEX RK schemes that preserve positivity for the classical single-species BGK equation have been recently developed in [HSZ18, GGHS22]. However, to our knowledge, these schemes cannot be applied directly to BGK equations with velocity-dependent collision frequencies.

Nevertheless, we derive some sufficient conditions on the time step Δt for positivity-preservation in the second-order schemes presented in Section 3.3.3 and 3.3.4.

Proposition 4.2.3 (Positivity of second-order splitting scheme). *The second-order splitting scheme presented in Section 3.3.3 together with the second-order space discretization described in Section 3.4 is positivity-preserving, provided that*

$$\Delta t \leq \min \left\{ \frac{2\Delta x}{3 \max |v^{(1)}|}, \frac{2}{\nu_{ii}^* + \nu_{ij}^*} \right\}. \tag{4.7}$$

Proof. The positivity during the transport steps $\mathcal{T}$ is guaranteed by the CFL condition

$$\Delta t \leq \frac{2\Delta x}{3 \max |v^{(1)}|}$$

because Heun's method (3.4) is an explicit SSP scheme (preserving any convex property) [PR05]. For the relaxation step, we derive the above condition by guaranteeing that the given functions G_i^* are positive. So we require

$$\begin{aligned} 0 \leq G_i^* &= f_i^* + \frac{\Delta t}{2}(\nu_{ii}^*(\mathcal{A}_{ii,\mathcal{T}_i}^* - f_i^*) + \nu_{ij}^*(\mathcal{A}_{ij,\mathcal{T}_i}^* - f_i^*)) \\ &= \left(1 - \frac{\Delta t}{2}(\nu_{ii}^* + \nu_{ij}^*)\right) f_i^* + \frac{\Delta t}{2}(\nu_{ii}^* \mathcal{A}_{ii,\mathcal{T}_i}^* + \nu_{ij}^* \mathcal{A}_{ij,\mathcal{T}_i}^*) \end{aligned}$$

for $i, j = 1, 2$. An obvious sufficient condition for positivity reads

$$\Delta t \leq \frac{2}{\nu_{ii}^* + \nu_{ij}^*}.$$

□

If ν_{ij}^* become large, the time step condition (4.7) can be very restrictive. So one may enforce instead

$$\Delta t \leq \frac{2 f_i^*}{\nu_{ii}^*(f_i^n - A_{ii,\mathcal{T}_{ii}}^*) + \nu_{ij}^*(f_i^* - \mathcal{A}_{ij,\mathcal{T}_i}^*)} \tag{4.8}$$

which is a milder, but still sufficient local condition. Large collision frequencies push each numerical kinetic distribution to the corresponding target function making the differences in the denominator small. It follows that condition (4.8) is not very restrictive.

We find similar results for the second-order IMEX RK scheme.

Proposition 4.2.4 (Positivity of second-order IMEX RK scheme). *For the space homogeneous case, the second-order IMEX RK scheme presented in Section 3.3.4 is positivity-preserving provided that*

$$\Delta t \leq \frac{1}{(1 - 2\gamma)(\nu_{ii}^{[1]} + \nu_{ij}^{[1]})} \tag{4.9}$$

for $i, j = 1, 2$.

Proof. The positivity of $f_i^{[1]}$ follows directly from its definition without any restriction on the time step. For the positivity of $f_i^{\ell+1} = f_i^{[2]}$, we require $G_i^{[2]} \geq 0$. Using the definition of $f_i^{[1]}$, we obtain

$$\begin{aligned} 0 \leq G_i^{[2]} &= f_i^\ell + \Delta t(1-\gamma)\left[\nu_{ii}^{[1]}\mathcal{A}_{ii,\tau_i}^{[1]} + \nu_{ij}^{[1]}\mathcal{A}_{ij,\tau_i}^{[1]} - (\nu_{ij}^{[1]} + \nu_{ij}^{[1]})f_i^{[1]}\right] \\ &= f_i^\ell\left[1 - \Delta t(1-\gamma)c_i^{[1]}(\nu_{ii}^{[1]} + \nu_{ij}^{[1]})\right] + \Delta t(1-\gamma)c_i^{[1]}\left[\nu_{ii}^{[1]}\mathcal{A}_{ii,\tau_i}^{[1]} + \nu_{ij}^{[1]}\mathcal{A}_{ij,\tau_i}^{[1]}\right]. \end{aligned}$$

Then, the most obvious sufficient condition for positivity reads

$$1 - \Delta t(1-\gamma)c_i^{[1]}(\nu_{ii}^{[1]} + \nu_{ij}^{[1]}) \geq 0 \quad \Longleftrightarrow \quad \Delta t \leq \frac{1}{(1-2\gamma)(\nu_{ii}^{[1]} + \nu_{ij}^{[1]})}.$$

□

The time step condition (4.9) can be very restrictive if $\nu_{ij}^{[1]}$ become large, especially for velocity-dependent collision frequencies. For this reason, one may instead enforce the milder (but still sufficient) local conditions

$$\Delta t \leq \frac{f_i^\ell}{(1-2\gamma)(\nu_{ii}^{[1]} + \nu_{ij}^{[1]})f_i^\ell - (1-\gamma)(\nu_{ii}^{[1]}\mathcal{A}_{ii,\tau_i}^{[1]} + \nu_{ij}^{[1]}\mathcal{A}_{ij,\tau_i}^{[1]})} \tag{4.10}$$

and

$$\Delta t \leq \frac{f_i^\ell}{(1-\gamma)\left[(\nu_{ii}^{[1]} + \nu_{ij}^{[1]})f_i^{[1]} - (\nu_{ii}^{[1]}\mathcal{A}_{ii,\tau_i}^{[1]} + \nu_{ij}^{[1]}\mathcal{A}_{ij,\tau_i}^{[1]})\right]}. \tag{4.11}$$

When the frequencies are large, the difference between each numerical kinetic distribution and its corresponding target function is to scale with the inverse of the frequency, in which case (4.10) and (4.11) are not very restrictive.

A distribution function of a fermion has the additional upper bound $f < 1$. The following proposition shows that our scheme preserves this property.

Proposition 4.2.5 (Boundedness for fermions). *If f_i represents the distribution function of a fermion with $f_i^\ell < 1$, any time discretization in Section 3.3.1–3.3.3 together with the space discretization described in Section 3.4 gives $f_i^{\ell+1} < 1$.*

Proof. We perform the proof for the first-oder time splitting method in Section 3.3.1. Let $f_i^\ell < 1$. The local equilibrium of a fermion is a Fermi-Dirac distribution function $\mathcal{F}$ for which $0 < \mathcal{F} < 1$ by definition. Hence, for the relaxation step it holds

$$f_i^* = c_i^\ell f_i^\ell + c_i^\ell \Delta t\,(\mathcal{F}_{ii}^* + \mathcal{F}_{ij}^*) < c_i^\ell + c_i^\ell 2\Delta t = 1 \tag{4.12}$$

because (3.51) becomes $c_i^\ell = \frac{1}{1+2\Delta t}$ in the quantum set-up. For the transport step (3.73), we obtain with the first-order fluxes

$$\begin{aligned} f_{i,k}^{\ell+1} &= \left(1 - \frac{\Delta t}{\Delta x}|v^{(1)}|\right) f_{i,k}^* + \frac{\Delta t}{\Delta x}|v^{(1)}| f_{i,k-\mathrm{sign}(v^{(1)})}^* \\ &\overset{(4.12)}{\leq} \left(1 - \frac{\Delta t}{\Delta x}|v^{(1)}|\right) + \frac{\Delta t}{\Delta x}|v^{(1)}| = 1. \end{aligned}$$

For the second-order fluxes, define $\sigma := \text{sign}(f^*_{i,k} - f^*_{i,k-1})$. Then one can show that

$$\phi_{k+\frac{1}{2}}(f^*_i) \leq \begin{cases} 0 & \text{if} \quad \sigma = -1 \\ f^*_{i,k+1} - f^*_{i,k} & \text{if} \quad \sigma = +1 \end{cases},$$

$$-\phi_{k-\frac{1}{2}}(f^*_i) \leq \begin{cases} f^*_{i,k-1} - f^*_{i,k} & \text{if} \quad \sigma = -1 \\ 0 & \text{if} \quad \sigma = +1 \end{cases}.$$

Hence,

$$\begin{aligned} f^{\ell+1}_{i,k} &\overset{(3.73)}{=} \left(1 - \frac{\Delta t}{\Delta x}|v^{(1)}|\right) f^*_{i,k} + \frac{\Delta t}{\Delta x}|v^{(1)}| f^*_{i,k-\text{sign}(v^{(1)})} \\ &\qquad + \frac{\Delta t}{\Delta x}\frac{|v^{(1)}|}{2}\left(\phi_{k+\frac{1}{2}}(f^*_i) - \phi_{k-\frac{1}{2}}(f^*_i)\right) \\ &\leq \left(1 - \frac{\Delta t}{\Delta x}|v^{(1)}|\right) f^*_{i,k} + \frac{\Delta t}{\Delta x}|v^{(1)}| f^*_{i,k-\text{sign}(v^{(1)})} \\ &\qquad + \frac{\Delta t}{\Delta x}\frac{|v^{(1)}|}{2} \begin{cases} (f^*_{i,k-1} - f^*_{i,k}) & \text{if} \quad \sigma = -1 \\ (f^*_{i,k+1} - f^*_{i,k}) & \text{if} \quad \sigma = +1 \end{cases} \\ &= \left(1 - \frac{3}{2}\frac{\Delta t}{\Delta x}|v^{(1)}|\right) f^*_{i,k} + \frac{\Delta t}{\Delta x}|v^{(1)}| f^*_{i,k-\text{sign}(v^{(1)})} \\ &\qquad + \frac{\Delta t}{\Delta x}\frac{|v^{(1)}|}{2} \begin{cases} f^*_{i,k-1} & \text{if} \quad \sigma = -1 \\ f^*_{i,k+1} & \text{if} \quad \sigma = +1 \end{cases} \\ &\overset{(4.12)}{\leq} 1. \end{aligned}$$

The proof for the time discretization techniques in Sections 3.3.2 and 3.3.3 works analogously. □

The fully-discrete scheme maintains the same properties regarding positivity and $f_i < 1$ for fermions when fulfilling the CFL condition with respect to the discrete velocities, see e.g. (4.13).

Corollary 4.2.6. *Propositions 4.2.1–4.2.5 all hold true after discretizing the velocity variable.*

In the following section, we summarize the time step restrictions of our scheme.

4.3 Time step restrictions

In order to guarantee stability and positivity, the advection in space introduces restrictions for the size of the time step resulting in a CFL condition. The Courant number C needs to be chosen such that

$$0 < C \leq \alpha.$$

It is $\alpha = 1$ for the first-order numerical fluxes and $\alpha = \frac{2}{3}$ for the second-order numerical fluxes. Considering only the convection parts, the time step size Δt for the update $t_\ell \to t_{\ell+1}$ is chosen to be

$$\Delta t = C \frac{\Delta x}{\max_{q_1} |v_{q_1}^{(1)}|}. \tag{4.13}$$

Additionally, we need to take care of the time step restrictions due to the preservation of positivity during relaxation steps. In our numerical tests, the time step Δt is set according to the CFL condition (4.13) by default. This guarantees positivity for the first-order time-stepping schemes. If positivity is violated for the second-order time-stepping schemes, we reduce the time step size according to (4.7) and (4.9), respectively. Thus we guarantee positivity while maintaining large time steps whenever possible. One could instead use the less restrictive local conditions in (4.8), (4.10) and (4.11), which requires additional iterations over the grid to find a global value for the time step. However, in practice, violations of positivity are rare and thus we use (4.7), respective (4.9) for simplicity.

In the following section, we address the conservation of mass, total momentum, and total energy for our scheme.

4.4 Conservation of mass, total momentum and total energy

To start with, we look at the relaxation and transport steps in the semi-discrete scheme, see Propositions 4.4.1 and 4.4.2. Then, we conclude with the fully-discrete scheme, see Theorem 4.4.4.

Proposition 4.4.1 (Conservation during relaxation process). *The relaxation step in the first-order splitting scheme presented in Section 3.3.1 satisfies the conservation laws*

$$\int \mathrm{a}_1^{(0)} f_1^* \, \mathrm{d}\mathbf{w} = \int \mathrm{a}_1^{(0)} f_1^\ell \, \mathrm{d}\mathbf{w}, \quad \int \mathrm{a}_2^{(0)} f_2^* \, \mathrm{d}\mathbf{w} = \int \mathrm{a}_2^{(0)} f_2^\ell \, \mathrm{d}\mathbf{w}, \tag{4.14}$$

$$\int \left(\mathbf{a}_1^{(1)} f_1^* + \mathbf{a}_2^{(1)} f_2^* \right) \mathrm{d}\mathbf{w} = \int \left(\mathbf{a}_1^{(1)} f_1^\ell + \mathbf{a}_2^{(1)} f_2^\ell \right) \mathrm{d}\mathbf{w}, \tag{4.15}$$

$$\int \left(\mathrm{a}_1^{(2)} f_1^* + \mathrm{a}_2^{(2)} f_2^* \right) \mathrm{d}\mathbf{w} = \int \left(\mathrm{a}_1^{(2)} f_1^\ell + \mathrm{a}_2^{(2)} f_2^\ell \right) \mathrm{d}\mathbf{w}. \tag{4.16}$$

Proof. We multiply the relaxation step (3.39) by $\mathbf{a}_i$, sum over $i = 1, 2$, and integrate with respect to $\mathbf{w}$. Sorting terms yields

$$\begin{aligned}
&\int (f_1^* \mathbf{a}_1 + f_2^* \mathbf{a}_2)\, \mathrm{d}\mathbf{w} - \int \left(f_1^\ell\, \mathbf{a}_1 + f_2^\ell\, \mathbf{a}_2\right) \mathrm{d}\mathbf{w} \\
&\overset{(3.39)}{=} \Delta t \Bigg[\int \left(c_1^\ell \nu_{11}^\ell A_{11,\tau_1}^* \mathbf{a}_1 + c_2^\ell \nu_{22}^\ell A_{22,\tau_2}^* \mathbf{a}_2 + c_1^\ell \nu_{12}^\ell A_{12,\tau_1}^* \mathbf{a}_1 + c_2^\ell \nu_{21}^\ell A_{21,\tau_2}^* \mathbf{a}_2\right) \mathrm{d}\mathbf{w} \\
&\qquad - \int \left[c_1^\ell \left(\nu_{11}^\ell + \nu_{12}^\ell\right) f_1^\ell\, \mathbf{a}_1 + c_2^\ell \left(\nu_{22}^\ell + \nu_{21}^\ell\right) f_2^\ell\, \mathbf{a}_2\right] \mathrm{d}\mathbf{w} \Bigg].
\end{aligned} \tag{4.17}$$

The RHS above corresponds to the first-order optimality conditions in (3.53). This term is identically zero as we minimize the corresponding functions in (3.58) and (3.59), which in turn proves the conservation statement (4.15) and (4.16). For the number of particles, we execute the above procedure for each species individually and obtain

$$\int f_i^* \mathrm{a}_i^{(0)}\, \mathrm{d}\mathbf{w} - \int f_i^\ell \mathrm{a}_i^{(0)}\, \mathrm{d}\mathbf{w} \tag{4.18}$$

$$\overset{(3.39)}{=} \Delta t \int \left(c_i^\ell \nu_{ii}^\ell A_{ii,\tau_i}^* \mathrm{a}_i^{(0)} + c_i^\ell \nu_{ij}^\ell A_{ij,\tau_i}^* \mathrm{a}_i^{(0)}\right) \mathrm{d}\mathbf{w} - \int c_i^\ell \left(\nu_{ii}^\ell + \nu_{ij}^\ell\right) f_i^\ell\, \mathrm{a}_i^{(0)}\, \mathrm{d}\mathbf{w} \tag{4.19}$$

$$= \Delta t \left[\partial_{\lambda_i^{(0)}} \varphi_i(\boldsymbol{\lambda}_i) + \partial_{\lambda_{ij}^{(0)}} \varphi(\boldsymbol{\lambda})\right] \tag{4.20}$$

which vanishes due to first-order optimality conditions on φ_i and φ being defined in (3.58) and (3.59). □

The conservation properties during the transport step are preserved which is inherited from the FV formulation.

Proposition 4.4.2 (Conservation during transport process). *For each $i = 1, 2$, the transport step in the first-order splitting scheme in Section 3.3.1 combined with the space discretization presented in Section 3.4 satisfies the conservation laws*

$$\sum_{k=0}^{K} \int \mathbf{a}_i f_{i,k}^{\ell+1}\, \mathrm{d}\mathbf{w} \Delta x = \sum_{k=0}^{K} \int \mathbf{a}_i f_{i,k}^*\, \mathrm{d}\mathbf{w} \Delta x$$

for periodic or zero boundary conditions.

Proof. For $i = 1, 2$, we multiply the transport step (3.41) by $\mathbf{a}_i$, integrate with respect to $\mathbf{w}$ and sum over all cell averages in x. The result is

$$\begin{aligned}
&\sum_{k=0}^{K} \int \mathbf{a}_i f_{i,k}^{\ell+1} \, \mathrm{d}\mathbf{w} \Delta x \\
&\qquad \overset{(3.41)}{=} \sum_{k=0}^{K} \int \mathbf{a}_i f_{i,k}^{*} \, \mathrm{d}\mathbf{w} \Delta x - \sum_{k=0}^{K} \int \Delta t \left(\mathscr{F}_{k+\frac{1}{2}}(f_i^*) - \mathscr{F}_{k-\frac{1}{2}}(f_i^*) \right) \mathbf{a}_i \, \mathrm{d}\mathbf{w} \\
&\qquad = \sum_{k=0}^{K} \int \mathbf{a}_i f_{i,k}^{*} \, \mathrm{d}\mathbf{w} \Delta x - \Delta t \, \boldsymbol{\Omega}
\end{aligned}$$

where the remnant of the telescoping sum

$$\boldsymbol{\Omega} = \int \mathbf{a}_i \mathscr{F}_{K+\frac{1}{2}}(f_i^*) \, \mathrm{d}\mathbf{w} - \int \mathbf{a}_i \mathscr{F}_{-\frac{1}{2}}(f_i^*) \, \mathrm{d}\mathbf{w}$$

vanishes for periodic or zero boundary conditions, e.g. $\mathscr{F}_{K+\frac{1}{2}}(f_i^*) = \mathscr{F}_{-\frac{1}{2}}(f_i^*)$ and $\mathscr{F}_{K+\frac{1}{2}}(f_i^*) = \mathscr{F}_{-\frac{1}{2}}(f_i^*) = 0$, respectively. □

Any time-stepping scheme presented in Sections 3.3.1–3.3.4 can be broken into relaxation and transport parts, each of which preserves the conservation of mass, total momentum, and total energy. As a result, we have the following.

Corollary 4.4.3. *For periodic or zero boundary conditions, any combination of temporal and space discretization presented in Sections 3.3 and 3.4, respectively, conserves mass, total momentum and total energy.*

Since the optimization algorithm is executed on the discrete-velocity grid, we now conclude with a statement on the fully-discrete scheme.

Theorem 4.4.4 (Conservation properties at the discrete level). *The scheme in Section 3.3.5 together with the space discretization and velocity discretization presented in Sections 3.4 and 3.5, respectively, satisfies the following conservation properties for* $\ell \geq 0$

$$\begin{aligned}
&\sum_{k,\mathbf{q}} \omega_{\mathbf{q}} \left(f_{1,k\mathbf{q}}^{\ell} \mathbf{a}_{1,\mathbf{q}} (\Delta w_1)^3 + f_{2,k\mathbf{q}}^{\ell} \mathbf{a}_{2,\mathbf{q}} (\Delta w_2)^3 \right) \Delta x \\
&\qquad = \sum_{k,\mathbf{q}} \omega_{\mathbf{q}} \left(f_{1,k\mathbf{q}}^{0} \mathbf{a}_{1,\mathbf{q}} (\Delta w_1)^3 + f_{2,k\mathbf{q}}^{0} \mathbf{a}_{2,\mathbf{q}} (\Delta w_2)^3 \right) \Delta x
\end{aligned} \tag{4.21}$$

with $\mathbf{a}_{i,\mathbf{q}}(\mathbf{v}) = m_i(1, \mathbf{v}_{\mathbf{q}}, |\mathbf{v}_{\mathbf{q}}|^2)^\top$, $\mathbf{a}_{i,\mathbf{q}}(\mathbf{p}) = (m_i, \mathbf{p}_{\mathbf{q}}, |\mathbf{p}_{\mathbf{q}}|^2/m_i)^\top$ *and* $f_{i,k\mathbf{q}}^{n} \approx f_{i,k}^{\ell}(\mathbf{v}_{\mathbf{q}})$.

We now come to the entropy behavior of our scheme.

4.5 Entropy inequality

We discuss the entropy behavior for the first-order scheme in Section 3.3.1. Both the relaxation and the transport step dissipate entropy; see Propositions 4.5.1 and 4.5.3. Additionally, we show in Proposition 4.5.2 that the minimal entropy is reached for the relaxation step if the distribution functions coincide with the corresponding target functions. These results rely on dissipative properties of the Backward Euler method and make use of the conservation properties at the semi-discrete level in a clever way.

Proposition 4.5.1 (Entropy inequality for the relaxation process in the first-order splitting scheme). *Let $h_\tau(f) = f\log f - f + \tau(1-\tau f)\log(1-\tau f) - \tau(1-\tau f)$. The relaxation step in the first-order splitting scheme in Section 3.3.1 fulfills the discrete entropy inequality*

$$\int h_{\tau_1}(f_1^*) + h_{\tau_2}(f_2^*)\,\mathrm{d}\mathbf{w} \le \int h_{\tau_1}(f_1^\ell) + h_{\tau_2}(f_2^\ell)\,\mathrm{d}\mathbf{w}. \tag{4.22}$$

Proof. By convexity

$$h_{\tau_i}(f_i^\ell) \ge h_{\tau_i}(f_i^*) + h'_{\tau_i}(f_i^*)(f_i^\ell - f_i^*). \tag{4.23}$$

The derivative

$$h'_\tau(f) = \log\frac{f}{1-\tau f}$$

is monotonically increasing for $f \ge 0$ ($\tau \in \{-1,0\}$) and $0 < f < 1$ ($\tau = +1$), respectively. This leads to the inequality

$$(h'(x) - h'(y))(y-x) \le 0 \tag{4.24}$$

for all $x, y \ge 0$ ($\tau \in \{-1,0\}$) and $0 < x, y < 1$ ($\tau = +1$), respectively. Moreover, because

$$h'_{\tau_i}(\mathcal{A}^*_{ij,\tau_i}) = \boldsymbol{\lambda}_{ij}\cdot\mathbf{a}(\mathbf{w})$$

it follows that

$$\int \nu_{ii}^\ell\, h'_{\tau_i}(\mathcal{A}^*_{ii,\tau_i})(\mathcal{A}^*_{ii,\tau_i} - f_i^*)\,\mathrm{d}\mathbf{w} = \boldsymbol{\lambda}_{ii}\cdot\int \nu_{ii}^\ell\,\mathbf{a}(\mathbf{w})(\mathcal{A}^*_{ii,\tau_i} - f_i^*)\,\mathrm{d}\mathbf{w} = 0. \tag{4.25}$$

Analogously for the inter-species terms,

$$\begin{aligned}
&\int \nu_{12}^\ell\, h'_{\tau_1}(\mathcal{A}^*_{12,\tau_1})(\mathcal{A}^*_{12,\tau_1} - f_1^*)\,\mathrm{d}\mathbf{w} + \int \nu_{21}^\ell\, h'_{\tau_2}(\mathcal{A}^*_{21,\tau_2})(\mathcal{A}^*_{21,\tau_2} - f_2^*)\,\mathrm{d}\mathbf{w} \\
&= \lambda_{12}^{(0)}\int \nu_{12}^\ell\,(\mathcal{A}^*_{12,\tau_1} - f_1^*)\,\mathrm{d}\mathbf{w} + \lambda_{21}^{(0)}\int \nu_{21}^\ell\,(\mathcal{A}^*_{21,\tau_2} - f_2^*)\,\mathrm{d}\mathbf{w} \\
&\quad + \begin{pmatrix}\boldsymbol{\lambda}^{(1)}\\ \lambda^{(2)}\end{pmatrix}\cdot\int\left(\nu_{12}^\ell\,(\mathcal{A}^*_{12,\tau_1} - f_1^*)\begin{pmatrix}\mathbf{a}_1^{(1)}(\mathbf{w})\\ a_1^{(2)}(\mathbf{w})\end{pmatrix} + \nu_{21}^\ell\,(\mathcal{A}^*_{21,\tau_2} - f_2^*)\begin{pmatrix}\mathbf{a}_2^{(1)}(\mathbf{w})\\ a_2^{(2)}(\mathbf{w})\end{pmatrix}\right)\mathrm{d}\mathbf{w} \\
&= 0.
\end{aligned} \tag{4.26}$$

The integrals (4.25) and (4.26) vanish because the conservation properties are satisfied at the semi-discrete level as well by construction of the scheme. Using the implicit step (3.39), i.e.

$$f_i^* - f_i^\ell = \Delta t \nu_{ii}^\ell (A_{ii,\tau_i}^* - f_i^*) + \Delta t \nu_{ij}^\ell (A_{ij,\tau_i}^* - f_i^*),$$

and the convexity of h_τ gives

$$\begin{aligned} h_{\tau_i}(f_i^*) - h_{\tau_i}(f_i^\ell) &\overset{(4.23)}{\leq} h'_{\tau_i}(f_i^*)(f_i^* - f_i^\ell) \\ &= \Delta t\, \nu_{ii}^\ell h'_{\tau_i}(f_i^*)(A_{ii,\tau_i}^* - f_i^*) + \Delta t \nu_{ij}^\ell h'_{\tau_i}(f_i^*)(A_{ij,\tau_i}^* - f_i^*). \end{aligned} \tag{4.27}$$

Thus after integrating (4.27) in $\mathbf{w}$ and making use of the zeros (4.25) and (4.26), we obtain

$$\begin{aligned} &\int h_{\tau_1}(f_1^*)\,d\mathbf{w} - \int h_{\tau_1}(f_1^\ell)\,d\mathbf{w} + \int h_{\tau_2}(f_2^*)\,d\mathbf{w} - \int h_{\tau_2}(f_2^\ell)\,d\mathbf{w} \\ &\leq \Delta t\, \nu_{11}^\ell \int [h'_{\tau_1}(f_1^*) - h'_{\tau_1}(\mathcal{A}_{11,\tau_1}^*)](\mathcal{A}_{11,\tau_1}^* - f_1^*)\,d\mathbf{w} \\ &\quad + \Delta t\, \nu_{22}^\ell \int [h'_{\tau_2}(f_2^*) - h'_{\tau_2}(\mathcal{A}_{22,\tau_2}^*)](\mathcal{A}_{22,\tau_2}^* - f_2^*)\,d\mathbf{w} \\ &\quad + \Delta t\, \nu_{12}^\ell \int [h'_{\tau_1}(f_1^*) - h'_{\tau_1}(\mathcal{A}_{12,\tau_1}^*)](\mathcal{A}_{12,\tau_1}^* - f_1^*)\,d\mathbf{w} \\ &\quad + \Delta t\, \nu_{21}^\ell \int [h'_{\tau_2}(f_2^*) - h'_{\tau_2}(\mathcal{A}_{21,\tau_2}^*)](\mathcal{A}_{21,\tau_2}^* - f_2^*)\,d\mathbf{w} \\ &\leq 0. \end{aligned} \tag{4.28}$$

The last inequality holds true because of (4.24). □

Proposition 4.5.2 (Entropy equality for the relaxation process in the first-order splitting scheme). *The inequality in Proposition 4.5.1 is an equality if and only if* $f_1^\ell = \mathcal{A}_{12,\tau_1}^\ell$ *and* $f_2^\ell = \mathcal{A}_{21,\tau_2}^\ell$. *In such cases* $f_1^* = \mathcal{A}_{12,\tau_1}^*$ *and* $f_2^* = \mathcal{A}_{21,\tau_2}^*$.

Proof. Suppose first that $f_1^\ell = \mathcal{A}_{12,\tau_1}^\ell$ and $f_2^\ell = \mathcal{A}_{21,\tau_2}^\ell$. Then according to the $\mathcal{H}$-Theorem 2.2.7/2.3.16,

$$\int h_{\tau_1}(f_1^\ell) + h_{\tau_2}(f_2^\ell)\,d\mathbf{w} \leq \int h_{\tau_1}(g_1) + h_{\tau_2}(g_2)\,d\mathbf{w} \tag{4.29}$$

for any measurable positive functions g_1 and g_2 such that

$$\int \mathrm{a}_1^{(0)} g_1 d\mathbf{v} = \int \mathrm{a}_1^{(0)} f_1^\ell\,d\mathbf{w}, \quad \int \mathrm{a}_2^{(0)} g_2\,d\mathbf{w} = \int \mathrm{a}_2^{(0)} f_2^\ell\,d\mathbf{w}, \tag{4.30}$$

$$\int \left(\mathbf{a}_1^{(1)}(\mathbf{w}) g_1 + \mathbf{a}_2^{(1)}(\mathbf{w}) g_2\right)\,d\mathbf{w} = \int \left(\mathbf{a}_1^{(1)}(\mathbf{w}) f_1^\ell + \mathbf{a}_2^{(1)}(\mathbf{w}) f_2^\ell\right)\,d\mathbf{w}, \tag{4.31}$$

$$\int \left(\mathrm{a}_1^{(2)}(\mathbf{w}) g_1 + \mathrm{a}_2^{(2)}(\mathbf{w}) g_2\right)\,d\mathbf{w} = \int \left(\mathrm{a}_1^{(2)}(\mathbf{w}) f_1^\ell + \mathrm{a}_2^{(2)}(\mathbf{w}) f_2^\ell\right)\,d\mathbf{w}. \tag{4.32}$$

These conditions are exactly those satisfied by f_1^* and f_2^* (cf. Theorem 4.4.1). Hence

$$\int h_{\tau_1}(f_1^\ell) + h_{\tau_2}(f_2^\ell)\,\mathrm{d}\mathbf{w} \le \int h_{\tau_1}(f_1^*) + h_{\tau_2}(f_2^*)\,\mathrm{d}\mathbf{w} \tag{4.33}$$

which shows that (4.22) is an equality. To show the converse statement, suppose that (4.22) holds as an equality. Then according to (4.28) $f_1^* = \mathcal{A}_{11,\tau_1}^* = \mathcal{A}_{12,\tau_1}^*$ and $f_2^* = \mathcal{A}_{21,\tau_2}^* = \mathcal{A}_{22,\tau_2}^*$. Therefore, by definition of $\mathcal{R}_i$ in (3.30), $\mathcal{R}_i(f_i^*, f_j^*) = 0$, which when plugged into (3.38), gives $f_1^\ell = f_1^*$ and $f_2^\ell = f_2^*$. □

Proposition 4.5.3 (Entropy inequality for the transport process in the first-order splitting scheme). *Let $h_\tau(f) = f\log f - f + \tau(1-\tau f)\log(1-\tau f) - \tau(1-\tau f)$. The transport step in the first-order splitting scheme in Section 3.3.1 combined with the first-order spatial discretization in Section 3.4 fulfills the discrete entropy inequality*

$$\sum_{k=0}^{K}\left\{\int h_{\tau_1}(f_{1,k}^{\ell+1}) + h_{\tau_2}(f_{2,k}^{\ell+1})\,\mathrm{d}\mathbf{w}\right\}\Delta x \le \sum_{k=0}^{K}\left\{\int h_{\tau_1}(f_{1,k}^*) + h_{\tau_2}(f_{2,k}^*)\,\mathrm{d}\mathbf{w}\right\}\Delta x \tag{4.34}$$

for periodic or zero boundary conditions, provided that

$$\Delta t \le \frac{\Delta x}{\max|v^{(1)}|}.$$

Proof. Using the notation $v^+ := \frac{v^{(1)}+|v^{(1)}|}{2}$ and $v^- := \frac{v^{(1)}-|v^{(1)}|}{2}$, we write the update formula of (3.41) with the first-order numerical fluxes as

$$\begin{aligned} f_{i,k}^{\ell+1} &= f_{i,k}^* - \frac{\Delta t}{\Delta x}\left(v^+ f_{i,k}^* + v^- f_{i,k+1}^* - v^+ f_{i,k-1}^* - v^- f_{i,k}^*\right) \\ &= \left(1 - \frac{\Delta t}{\Delta x}|v^{(1)}|\right) f_{i,k}^* - \frac{\Delta t}{\Delta x} v^- f_{i,k+1}^* + \frac{\Delta t}{\Delta x} v^+ f_{i,k-1}^*. \end{aligned}$$

Clearly, if the CFL condition is fulfilled, then $f_{i,k}^{\ell+1}$ is a convex linear combination of $f_{i,k}^*$, $f_{i,k-1}^*$, and $f_{i,k+1}^*$. Thus by the convexity of h_τ, for each $v^{(1)}$,

$$\begin{aligned} &\sum_{k=0}^{K} h_{\tau_i}(f_{i,k}^{\ell+1})\Delta x \\ &\quad \le \sum_k \left[\left(1 - \frac{\Delta t}{\Delta x}|v^{(1)}|\right) h_{\tau_i}(f_{i,k}^*) - \frac{\Delta t}{\Delta x} v^- h_{\tau_i}(f_{i,k+1}^*) + \frac{\Delta t}{\Delta x} v^+ h_{\tau_i}(f_{i,k-1}^*)\right]\Delta x \\ &\quad = \sum_k h_{\tau_i}(f_{i,k}^*)\Delta x + \Delta t\,\Omega \end{aligned} \tag{4.35}$$

where the boundary term

$$\Omega = v^- f_{i,0}^* \log(f_{i,0}^*) - v^- f_{i,K+1}^* \log(f_{i,K+1}^*) - v^+ f_{i,K}^* \log(f_{i,K}^*) + v^+ f_{i,-1}^* \log(f_{i,-1}^*)$$

is the only remnant of the telescoping sum and vanishes for periodic or zero boundary conditions. Thus summation over i and integration of (4.35) with respect to $\mathbf{w}$ yields the entropy inequality in (4.34).

□

Combining the two results above gives the following:

Corollary 4.5.4. *For periodic or zero boundary conditions, the first-order splitting scheme 3.3.1 combined with the first-order numerical fluxes in Section 3.4 fulfills the discrete entropy inequality*

$$\sum_{k=0}^{K}\left\{\int h_{\tau_1}(f_{1,k}^{\ell+1}) + h_{\tau_2}(f_{2,k}^{\ell+1})\,\mathrm{d}\mathbf{w}\right\}\Delta x \leq \sum_{k=0}^{K}\left\{\int h_{\tau_1}(f_{1,k}^{*}) + h_{\tau_2}(f_{2,k}^{*})\,\mathrm{d}\mathbf{w}\right\}\Delta x$$

provided that

$$\Delta t \leq \frac{\Delta x}{\max|v^{(1)}|}.$$

Finally, we conclude with a statement on the fully-discrete scheme.

Theorem 4.5.5 (Entropy behavior of the first-order splitting scheme). *Propositions 4.5.1–4.5.3 all hold true after replacing continuous integrals by their respective quadratures.*

Chapter 5

Numerical Results

In the previous chapters, we have introduced fluid models and an adequate scheme for multi-species BGK-type equations. We now apply the scheme to the equations.

Firstly, we perform tests to verify the order of the presented scheme in Section 5.1.

In Section 5.2, we provide different test cases for the multi-species BGK model with constant collision frequency where we compare the results with analytic values or simulations from the literature.

In Section 5.3, we focus on velocity-dependent collision frequencies. We illustrate the behavior and properties of the scheme, followed by test cases inspired by physical experiments where we emphasize the effect of velocity-dependent collision frequencies by comparing the results with simulations using constant collision frequencies.

Afterwards, we present numerical results for the quantum multi-species model in Section 5.4. We verify the properties of the scheme and compare the analytical decay rates with the results from the numerical simulations. We conclude with a test case inspired by physical experiments.

5.1 Proof of order

In Section 5.1.1, we examine the order in space and time discretization, leaving out the velocity variable for reasons of computational costs. In Section 5.1.2, we leave out the space variable and run a relaxation test case which uses a discretization in time and velocity.

5.1.1 Transport equation

We consider the simple transport equation

$$\partial_t f(x,t) + v\,\partial_x f(x,t) = 0$$

where $v = 2$ in the domain $x \in [-1, 1]$ with periodic boundaries. The function f is initialized by

$$f(x,0) = 0.5\,\sin(\pi x) + 1.$$

This problem has the unique solution

$$f(x,t) = 0.5\,\sin(\pi(x-2t)) + 1, \tag{5.1}$$

see Section 2.1.2.

For the simulations, we refine the grid more and more. The coarsest grid has 10 equally spaced cells in x, and we compute one time step where Δt is set according to the CFL condition (3.74). The next coarsest grid has 20 equally spaced cells in x, and so on.

In Figure 5.1, the L_1-error of the numerical solutions compared to the analytical solution (5.1) is presented in a log-log plot. The first-order schemes from Sections 3.3.1 and 3.3.2 use the Forward Euler time discretization (3.2) combined with the first-order numerical fluxes from Section 3.4.1. This combination reproduces a good first-order convergence (left plot).

The nominally second-order schemes use the numerical fluxes with flux limiter (3.71) from Section 3.4.1. The splitting scheme from Section 3.3.3 implements Heun's method (3.4), whereas the IMEX RK from Section 3.3.4 uses the time discretization ARS-ex given in (3.6). Both methods give similar results. They do not show a good second-order convergence, but they definitely perform better than first order (right plot). This is typical for positivity-preserving flux limiters as they introduce numerical dissipation.

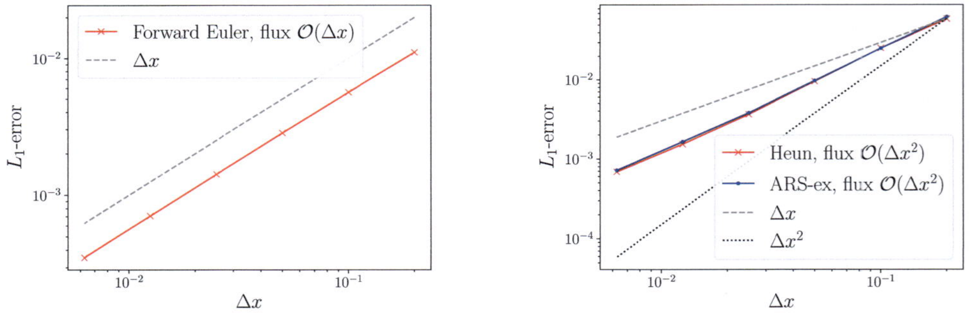

Figure 5.1: Numerical results for the test case 5.1.1. The L_1-error of the numerical solution compared to the analytical solution (5.1) is illustrated in a log-log plot. The Forward Euler time discretization (3.2) combined with the first-order numerical fluxes from Section 3.4.1 shows a good first-order convergence. The nominally second-order schemes (using a combination of Heun's method (3.4), respective ARS-ex (3.6) and the numerical fluxes with flux limiter from Section 3.4.1) converge faster than first order, but they do not quite reach a second-order convergence.

5.1.2 Relaxation test case

We consider a space homogeneous relaxation process between two species for the velocity-independent collision frequencies $\nu_{ij} = n_j$. Initially, we set the distribution functions to Maxwellians $f_i = \mathcal{M}[n_i, \mathbf{u}_i, T_i, m_i]$ with

$$
\begin{aligned}
m_1 &= 1.0, \quad & n_1 &= 1.0, \quad & \mathbf{u}_1 &= (1.0, 0, 0)^\top, \quad & T_1 &= 2.0, \\
m_2 &= 2.0, \quad & n_2 &= 1.2, \quad & \mathbf{u}_2 &= (0.5, 0, 0)^\top, \quad & T_2 &= 1.0.
\end{aligned}
$$

Again, we refine the grid more and more. The coarsest velocity grid has 12^3 nodes, and we set the time step to $\Delta t = 0.02$. The grid and time step are refined by factors of 2.

We cannot solve the equations analytically such that we use a numerical solution which is computed on a fine grid as reference solution. This grid uses 192^3 velocity nodes and a time step $\Delta t = 0.00125$. The reference solution is obtained by the second-order IMEX RK scheme from Section 3.3.4.

In Figure 5.2, we show the L_1-error of the numerical solution compared to the reference solution in a log-log plot. The first-order schemes from Sections 3.3.1 and 3.3.2 use the Backward Euler time discretization (3.8) which reproduces a good first-order convergence (left plot).

The second-order splitting scheme from Section 3.3.3 implements the implicit trapezoidal method (3.10). It gives a second-order convergence (middle plot). The IMEX RK from Section 3.3.4 uses the time discretization ARS-im given in (3.12) and shows a good second-order convergence (right plot). Moreover, the second-order schemes are more accurate as the L_1-error becomes smaller.

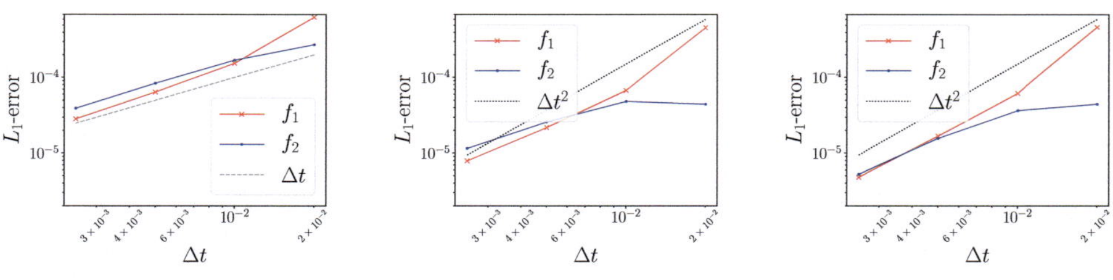

(a) Backward Euler method **(b)** Trapezoidal method **(c)** ARS-im method

Figure 5.2: Numerical results for the test case 5.1.2. The L_1-error of the numerical solution compared to the reference solution is illustrated in a log-log plot. The Backward Euler time discretization (3.8) shows a good first-order convergence. The implicit trapezoidal method (3.10) and the method ARS-im (3.12) show a second-order convergence and a higher accuracy.

We have verified the order of the individual explicit and implicit parts of the schemes. We rely on the construction of the IMEX schemes such that we do not perform any further tests regarding the order of the schemes. Instead we are interested in more physically relevant test cases.

5.2 Numerical results for the classic multi-species BGK model with constant collision frequencies

We present several numerical tests for the KPP model from Section 2.1.4. We verify the decay rates for the mean velocities and temperatures in Section 5.2.1. In Sections 5.2.2 and 5.2.3, we rerun test cases from the literature being physically motivated.

In the following test cases, we define δ, γ, α of the KPP model by assuming $\mathbf{u}_{12} = \mathbf{u}_{21}$ and $T_{12} = T_{21}$. This leads to

$$\delta = \frac{\varepsilon m_1}{m_1 \varepsilon + m_2}, \qquad \alpha = \frac{\varepsilon}{1+\varepsilon}, \qquad \gamma = \frac{\varepsilon m_1 (1-\delta) 2\delta}{3(1+\varepsilon)}, \qquad \varepsilon = \frac{m_2}{m_1}.$$

5.2.1 Decay rates

This test case is inspired from [CKP20]. We consider a space homogeneous relaxation process between two species for the velocity-independent collision frequencies

$$\nu_{ij} = 20 n_j.$$

Hence, analytical decay rates are given in (2.63) and (2.65). Initially, we set the distribution functions to Maxwellians $f_i = \mathcal{M}[n_i, \mathbf{u}_i, T_i, m_i]$ with

$$\begin{aligned} m_1 &= 1.5, \quad & n_1 &= 1.2, \quad & \mathbf{u}_1 &= (0.1, 0, 0)^\top, \quad & T_1 &= 0.1, \\ m_2 &= 1.0, \quad & n_2 &= 1.0, \quad & \mathbf{u}_2 &= (0.5, 0, 0)^\top, \quad & T_2 &= 1.0. \end{aligned}$$

For the simulation, we use a velocity grid with 48^3 nodes.

In Figure 5.3, we illustrate the results for different time steps and for the time discretization methods from Sections 3.3.1–3.3.4. The mean velocities converge exponentially fast to a common value which is clearly visible by the logarithmic plotting scale. The same holds true for the temperature, however, the decay rate needs not to be exponential in general. (It is a sum of exponential functions.) All schemes hit the analytic decay rates very well for a small time step. However, larger time steps amplify the numerical errors which makes the schemes less accurate affecting especially the first-order schemes.

5.2.2 Hydrogen-Carbon test case 1

We want to reproduce the result of a more realistic test case, firstly presented in [HHM17a] and inspired from plasma physics. It considers the space homogeneous mixture of hydrogen (species 1) and carbon (species 2).

First-order splitting scheme from Section 3.3.1:

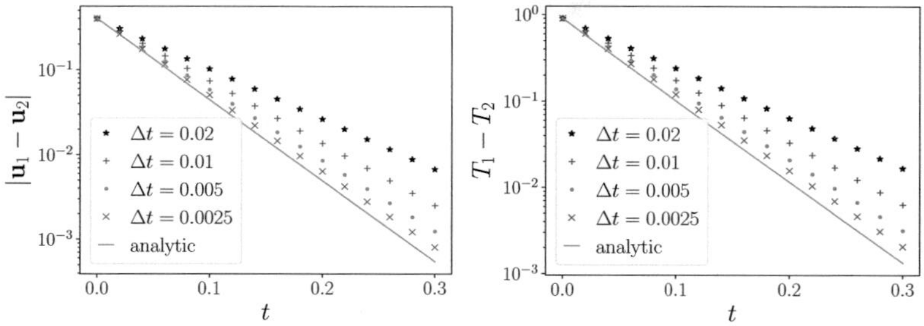

First-order IMEX RK scheme from Section 3.3.2:

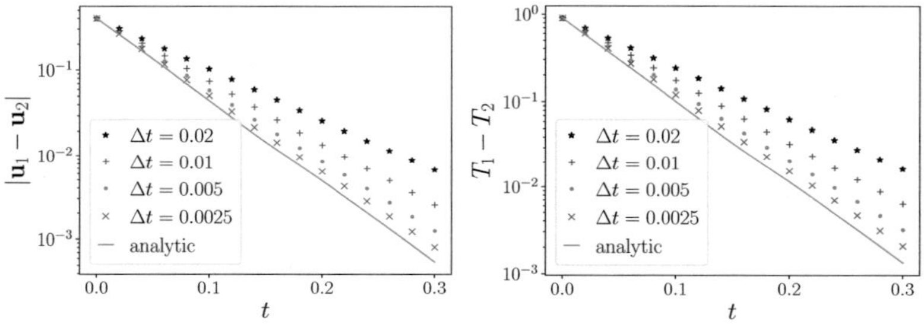

Second-order splitting scheme from Section 3.3.3:

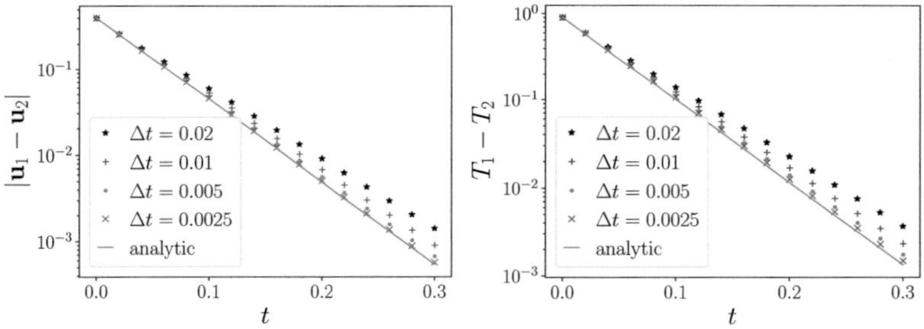

Second-order IMEX RK scheme from Section 3.3.4:

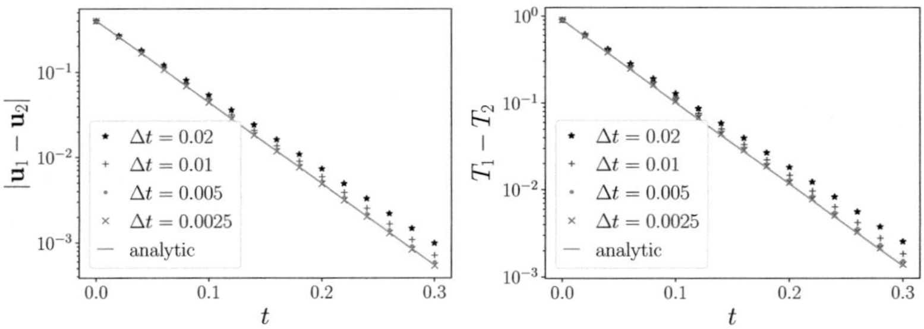

Figure 5.3: The mean velocities and temperatures of species 1 and 2 converge to a common value. The smaller the time step, the better match the numerical decay rates the analytic ones. The second-order schemes are more accurate which can be clearly seen.

We use the (velocity-independent) collision frequencies (2.106) and (2.107), see Section 2.2.3, i.e.

$$\nu_{ij}^M = \frac{128}{3}\frac{\pi^2}{(2\pi)^{3/2}}\frac{n_j}{m_i}(Z_j Z_i e^2)^2\frac{\sqrt{m_i m_j}(m_i+m_j)}{(m_i T_j + m_j T_i)^{3/2}}\,\Psi\left(\gamma_{ij}\right), \tag{5.2}$$

$$\nu_{ij}^T = \frac{256}{3}\frac{\pi^2}{(2\pi)^{3/2}}n_j(Z_j Z_i e^2)^2\frac{\sqrt{m_i m_j}}{(m_i T_j + m_j T_i)^{3/2}}\,\Psi\left(\gamma_{ij}\right). \tag{5.3}$$

The masses and charge numbers of the species are given by

$$\begin{aligned} m_1 &= 1.993 \cdot 10^{-23}\ \mathrm{g}, & m_2 &= 1.661 \cdot 10^{-24}\ \mathrm{g}, \\ Z_1 &= 6, & Z_2 &= 1. \end{aligned} \tag{5.4}$$

Initially, the distribution functions are Maxwellians $f_i = \mathcal{M}[n_i, \mathbf{u}_i, T_i, m_i]$ with

$$\begin{aligned} n_1 &= 10^{23}\ \mathrm{cm{-}3}, & n_2 &= 10^{23}\ \mathrm{cm}^{-3}, \\ \mathbf{u}_1 &= (1.26 \cdot 10^5,\, 0,\, 0)^\top\,\frac{\mathrm{cm}}{\mathrm{s}}, & \mathbf{u}_2 &= (0,\, 0,\, 0)^\top\,\frac{\mathrm{cm}}{\mathrm{s}}, \\ T_1 &= 10\ \mathrm{eV}, & T_2 &= 12\ \mathrm{eV}. \end{aligned} \tag{5.5}$$

The velocity grid takes 48^3 nodes, and we use the second-order IMEX RK scheme from Section 3.3.4 with time step $\Delta t = 0.1$ fs.

Remark 5.2.1 (Units in numerical test cases). *Incorporating units in a code is challenging, especially if particles with very small masses are considered. To overcome this problem, we normalize the units by*

$$\mathrm{cm} \to 1.5345 \cdot 10^7, \quad \mathrm{g} \to 6.02047 \cdot 10^{23}, \quad \mathrm{s} \to 1.50658 \cdot 10^{14},$$

resulting e.g. in the masses $m_1 = 12$ and $m_2 = 1$. We obtain initial values which can be handled better by numerical calculations. The values are reconverted in the CGS system when plotted.

This procedure is executed for any test case carrying CGS units.

In Figures 5.4 and 5.5, we illustrate the convergence of the mean velocities and temperatures to a common value which is already predetermined by the initial data and the mixture quantities (2.8) and (2.9). Comparing with [HHM17a], the results are in good agreement; however, we observe a slightly slower relaxation rate. We expect that this is due to the collision frequencies. Even though we took the formulas from [HHM17a], some details have been left open. Our choices, given in Section 2.2.3, may influence the result in the given way.

5.2.3 Sulfur-Fluorine-Electrons test case

For a 3-species test case, we reproduce the space homogeneous example given in [HHM17a]. In the following, the index S refers to sulfur ions, the index F refers to fluorine ions, and the index e refers to electrons.

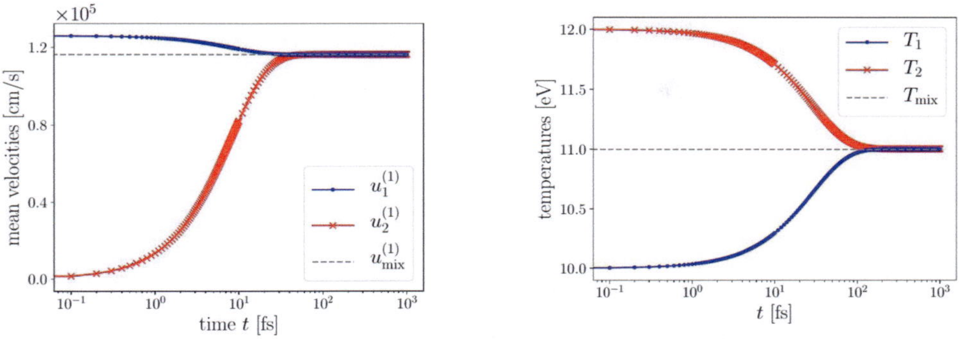

(a) mean velocities; using ν_{ij}^M

(b) temperatures; using ν_{ij}^M

Figure 5.4: Evolution of the species' mean velocities and temperatures for the Hydrogen-Carbon test case in Section 5.2.2 using the constant collision frequencies (5.2). The quantities hit the mixture quantities (2.8) and (2.9) as expected.

For ion-ion interactions, we use the collision frequencies (5.2), respective (5.3) from the previous numerical example. The collision frequencies which encounter an interaction with electrons are given in (2.109) and (2.110), respectively. For convenience, we restate them here:

$$\nu_{ij}^M = \frac{8}{3}\sqrt{2\pi}\, n_j \sqrt{\frac{m_j}{m_i}} \frac{m_i + m_j}{(m_i T_j + m_j T_i)^{3/2}} (Z_i Z_j e^2)^2 \log \Lambda_{\mathrm{GMS6}}, \tag{5.6}$$

$$\nu_{ij}^T = \frac{16}{3}\sqrt{2\pi}\, n_j \frac{\sqrt{m_i m_j}}{(m_i T_j + m_j T_i)^{3/2}} (Z_i Z_j e^2)^2 \log \Lambda_{\mathrm{GMS6}}. \tag{5.7}$$

Details can be found in Section 2.2.3.

The species' masses and charge numbers are

$$m_S = 32.07\mathrm{u} - 11 m_e, \quad m_F = 19\mathrm{u} - 7 m_e, \quad m_e = m_e,$$
$$Z_S = 11, \quad Z_F = 7, \quad Z_e = -1$$

with the atomic mass $\mathrm{u} = 1.6605 \cdot 10^{-24}$ g. Initially, we assume Maxwellian distributions $f_i = \mathcal{M}[n_i, \mathbf{u}_i, T_i, m_i]$ with

$$n_S = 10^{19}\,\mathrm{cm}^{-3}, \quad n_F = 6 \cdot 10^{19}\,\mathrm{cm}^{-3}, \quad n_e = 53 \cdot 10^{19}\,\mathrm{cm}^{-3},$$
$$\mathbf{u}_S = \mathbf{u}_F = \mathbf{u}_e = 0\,\frac{\mathrm{cm}}{\mathrm{s}},$$
$$T_S = T_F = 15\,\mathrm{eV}, \quad T_e = 100\,\mathrm{eV}.$$

We use velocity grids with 48^3 nodes for each species, and we use the second-order IMEX RK scheme from Section 3.3.4 with time step $\Delta t = 100$ fs.

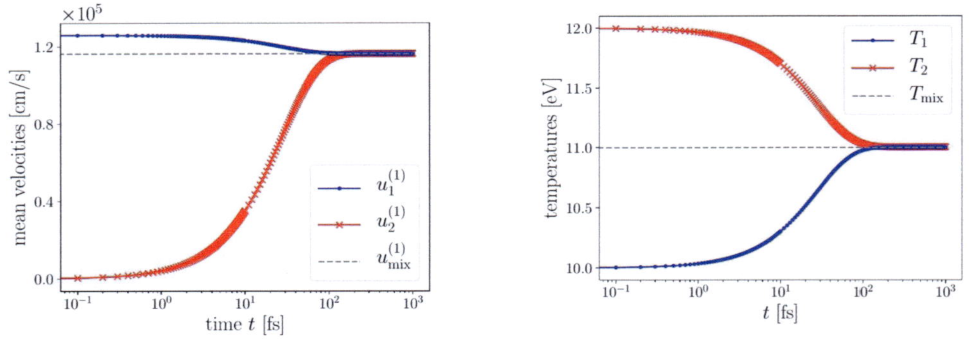

(a) mean velocities; using ν_{ij}^T

(b) temperatures; using ν_{ij}^T

Figure 5.5: Evolution of the species' mean velocities and temperatures for the Hydrogen-Carbon test case in Section 5.2.2 using the constant collision frequencies (5.3). The quantities hit the mixture quantities (2.8) and (2.9) as expected.

The species are initialized with vanishing mean velocity, and the mean velocities stay zero. We give the evolution of the temperatures in Figure 5.6 which converge to the mixture temperature (2.9). As in the previous test case, the relaxation rate is slower compared to the results in [HHM17a].

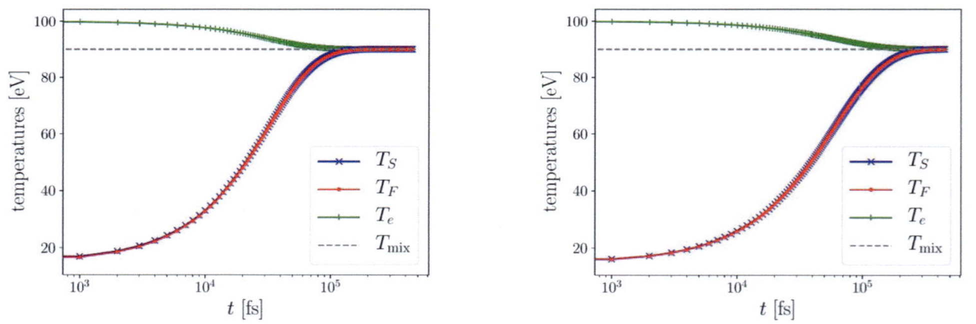

(a) temperatures; using ν_{ij}^M

(b) temperature; using ν_{ij}^T

Figure 5.6: Evolution of the species' temperatures for the Sulfur-Fluorine-electrons test case in Section 5.2.3 using the constant collision frequencies (5.2) and (5.6), respective (5.3) and (5.7). The quantities hit the mixture temperature (2.9) as expected.

For the following section, we increase the complexity of the underlying model by considering collision frequencies which are dependent on $\mathbf{v}$.

5.3 Numerical results for the classic multi-species BGK model with velocity-dependent collision frequencies

We perform numerical experiments in order to illustrate the effect of velocity-dependent collision frequencies in the multi-species BGK model from Section 2.2. Most of this section can also be found in [HHK$^+$22]. To start with, we examine various space homogeneous set-ups. We illustrate the properties at the discrete level in Section 5.3.1. A physical more relevant test case is provided in Section 5.3.2.

Afterwards, we run several shock tube problems for the full equations (2.69). In Section 5.3.3, we compare the well-known Sod problem in the hydrodynamic limit when approximated by BGK equations with different collision frequencies. We further examine the different behavior of constant in $\mathbf{v}$ and velocity-dependent collision frequencies for various shock wave problems in Sections 5.3.4 and 5.3.5. We conclude with the interpenetration of hydrogen and helium particles in Sections 5.3.6 and 5.3.7, where the use of different collision frequencies displays a significantly different behavior of the hydrogen species.

5.3.1 Illustrative toy problem

The purpose of this experiment is to illustrate basic properties of the multi-species BGK model with velocity-dependent collision frequency from Section 2.2. We solve the spatially homogeneous version of (2.69) for species with masses $m_1 = 1$ and $m_2 = 1.5$. The initial distribution functions (see Figure 5.7a) are given by

$$f_i(\mathbf{v}, t = 0) = 0.1 \cdot m_i^{27} \cdot \exp\left(- \frac{0.01}{(0.75/m_i)^{10} - |\mathbf{v} - \mathbf{u}_i(0)|_1^{10}} \right),$$

with $\mathbf{u}_1(0) = (0.1, 0, 0)^\top$ and $\mathbf{u}_2(0) = (-0.1, 0, 0)^\top$. The parameter choices here are not physical; rather they are chosen to yield an initial distribution with a particular shape that makes the relaxation easier to visualize. With this initialization, the mixture mean velocity and mixture temperature, defined in Definition 2.1.16, have numerical values

$$\mathbf{u}_{\text{mix}} = 0.0322 \quad \text{and} \quad T_{\text{mix}} = 0.0487.$$

According to Proposition 2.1.33, these values stay constant in time. The collision frequencies take the form

$$\nu_{ij}(x, \mathbf{v}, t) = \frac{10\, n_j}{\delta_{ij} + |\mathbf{v} - \mathbf{u}_{\text{mix}}|^3}$$

with the regularization parameter $\delta_{ij} = 0.1 \cdot (\Delta v_{ij})^3$ where $\Delta v_{ij} = \frac{1}{4}\sqrt{T_{\text{mix}}/(2\mu_{ij})}$, and the reduced mass is $\mu_{ij} = m_i m_j/(m_i + m_j)$.

The simulation is run using a velocity grid with 48^3 nodes and the first-order temporal splitting scheme from Section 3.3.1 with time step $\Delta t = 0.01$. As demonstrated in Chapter 4, this scheme maintains positivity, conservation, and entropy dissipation properties of the continuous model.

In Figure 5.7, we plot the kinetic distributions at several different times and observe convergence to their respective equilibria. It is easy to see that the convergence to equilibrium is much faster in the center than near the tails of the distribution functions. This is a consequence of the fact that the velocity-dependent collision frequency amplifies the relaxation process for small relative velocities.

In Figure 5.8, we show convergence of the mean velocities and temperatures to their equilibrium values, given by the mixture values in (2.8) and (2.9).

In Figure 5.9, we show the evolution of the entropy and the entropy dissipation. As expected, the entropy decays monotonically.

In Figure 5.10, we demonstrate conservation properties. The scheme conserves mass, total momentum and total energy up to numerical oscillations of the order 10^{-15} or less.

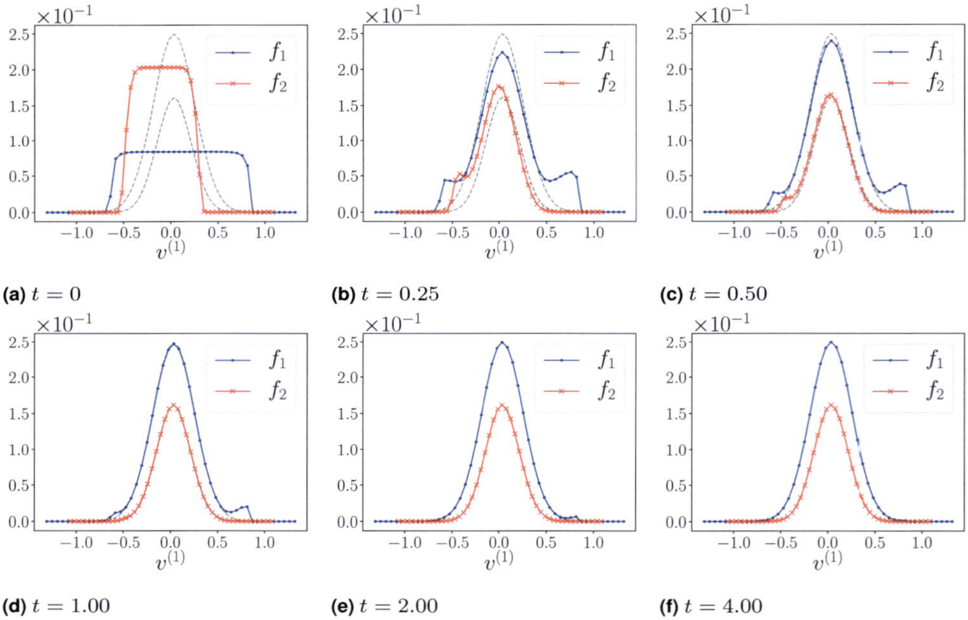

(a) $t = 0$ (b) $t = 0.25$ (c) $t = 0.50$

(d) $t = 1.00$ (e) $t = 2.00$ (f) $t = 4.00$

Figure 5.7: Relaxation of the distribution functions to Maxwellians for the test case in Section 5.3.1. We fix $v^{(2)} = v^{(3)} = 0$ and plot $f_i(v^{(1)}, v^{(2)} = 0, v^{(3)} = 0, t)$ at times t. At time progress, the two distribution functions converge to Maxwellians centered around a common mean velocity with a width according to their common temperature divided by mass. For reference, these Maxwellians are shown by dotted gray lines.

5.3.2 Hydrogen-Carbon test case 2

In this test case, we explore the effects of the velocity-dependent frequencies on the relaxation behavior of a multi-species problem in a more physically relevant setting,

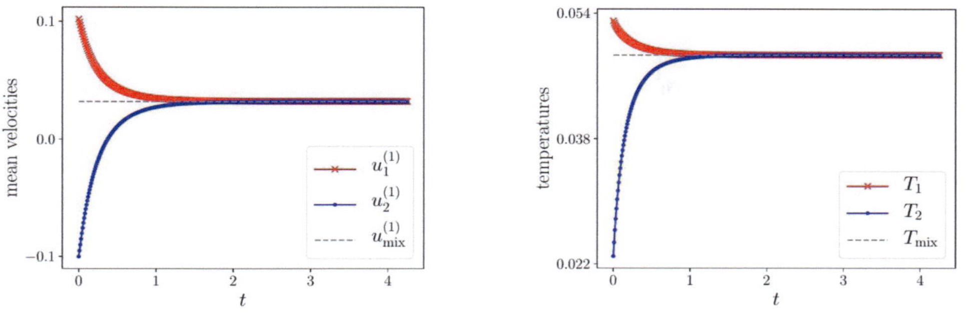

(a) convergence of mean velocities

(b) convergence of temperatures

Figure 5.8: Convergence of mean velocities and temperatures for the test case in Section 5.3.1. In each plot, the dotted line denotes the mixture values, given in (2.8) and (2.9).

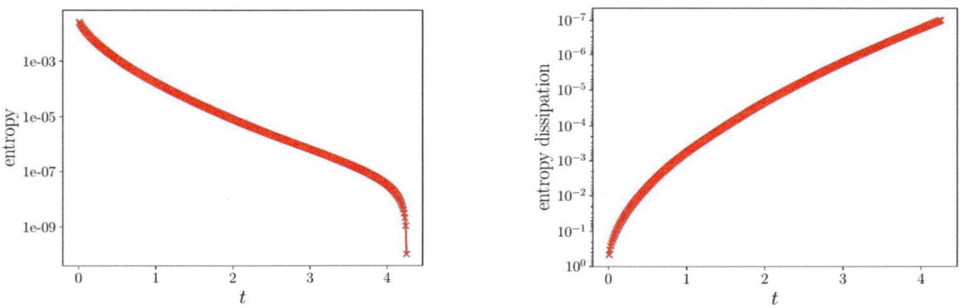

(a) entropy

(b) entropy dissipation

Figure 5.9: Entropy and entropy dissipation for the test case in Section 5.3.1. As predicted by the theory, the entropy decays monotonically.

with dimensional formulas given in the CGS unit system.

The collision frequencies which are used in this test case are derived in Section 2.2.3. For convenience, we give again

$$\nu_{ij}(\mathbf{v}) = 4\pi n_j \left(\frac{Z_i Z_j e^2}{2\mu_{ij}}\right)^2 \frac{1}{\delta_{ij} + |\mathbf{v} - \mathbf{u}_{\mathrm{mix}}|^3} \log \Lambda_{ij}, \tag{5.8}$$

where μ_{ij} is the reduced mass, $Z_i e$ and $Z_j e$ are the charges of the two particles and $\log \Lambda_{ij}$ is the Coulomb logarithm. Further details can be found in Section 2.2.3. The small regularization parameter $\delta_{ij} > 0$ in the denominator of (5.8) avoids a singularity at zero relative velocity. For the numerical experiments one needs to ensure that δ_{ij} is much smaller than $|\mathbf{v} - \mathbf{u}_{\mathrm{mix}}|^3$, and thus we set $\delta_{ij} = 0.1 \cdot (\Delta v_{ij})^3$, where $\Delta v_{ij} = \frac{1}{4}\sqrt{k_B T_{\mathrm{mix}}/(2\mu_{ij})}$. This choice ensures that ν_{12} and ν_{21} are symmetric up to the densities.

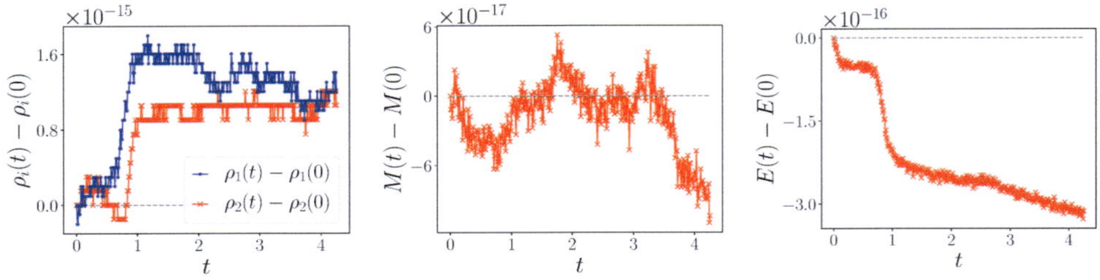

(a) species mass **(b)** total momentum **(c)** total energy

Figure 5.10: Global conservation properties for the test case in Section 5.3.1. The mass densities of each species, the total momentum (M) and total energy (E) have small oscillations of the order of 10^{-15} or less.

For comparison, we consider three velocity-independent collision frequencies that are often used as simpler alternatives to (5.8).

1. Replacing $|\mathbf{v} - \mathbf{u}_{\text{mix}}|$ by the thermal velocity $v_T = \sqrt{k_B T_{\text{mix}}/(2\mu_{ij})}$ gives

$$\tilde{\nu}_{ij} = 4\pi n_j \left(\frac{Z_i Z_j e^2}{2\mu_{ij}} \right)^2 \frac{1}{\delta_{ij} + v_T^3} \log \Lambda_{ij}. \tag{5.9}$$

2. Replacing $|\mathbf{v} - \mathbf{u}_{\text{mix}}|^3$ by the weighted average

$$\hat{v}^3 = \frac{\int |\mathbf{v} - \mathbf{u}_{\text{mix}}|^3 \mathcal{M}_{ij}(\mathbf{v})\,\mathrm{d}\mathbf{v}}{\int \mathcal{M}_{ij}(\mathbf{v})\,\mathrm{d}\mathbf{v}}, \tag{5.10}$$

   where $\mathcal{M}_{ij}(\mathbf{v}) = \mathcal{M}[n_i, \mathbf{u}_{\text{mix}}, T_{\text{mix}}, \mu_{ij}]$, gives

$$\hat{\nu}_{ij} = 4\pi n_j \left(\frac{Z_i Z_j e^2}{2\mu_{ij}} \right)^2 \frac{1}{\delta_{ij} + \hat{v}^3} \log \Lambda_{ij}. \tag{5.11}$$

3. Computing a weighted average of $\nu_{ij}(\mathbf{v})$ directly gives

$$\bar{\nu}_{ij} = \frac{\int \nu_{ij}(\mathbf{v}) \mathcal{M}_{ij}(\mathbf{v})\,\mathrm{d}\mathbf{v}}{\int \mathcal{M}_{ij}(\mathbf{v})\,\mathrm{d}\mathbf{v}}. \tag{5.12}$$

While the first option above is convenient and more common in applications [SM16], it is somewhat arbitrary. The second and third options, on the other hand, provide a more consistent normalization. According to Proposition 2.1.33, the collision frequencies stay constant in time because the problem is spatially homogeneous. For purposes of illustration, we plot them in Figure 5.11.

We consider relaxation between carbon (species 1) and hydrogen (species 2), with masses and charge numbers

$$\begin{aligned} m_1 &= 1.993 \cdot 10^{-23}\,\text{g}, & m_2 &= 1.661 \cdot 10^{-24}\,\text{g}, \\ Z_1 &= 6, & Z_2 &= 1. \end{aligned}$$

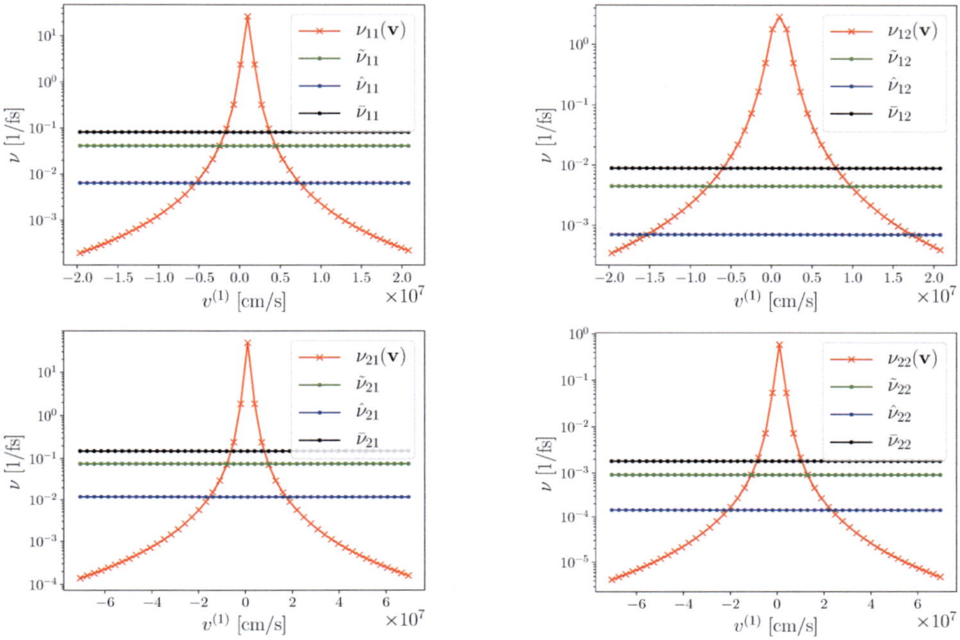

Figure 5.11: Collision frequencies given in (5.8), (5.9), (5.11) and (5.12) along the line $v^{(2)} = v^{(3)} = 0$. The largest constant values for $\bar{\nu}$ correspond to the fastest relaxation process, see Figure 5.12.

Initially, the distribution functions are Maxwellians $f_i = \mathcal{M}[n_i, \mathbf{u}_i, T_i, m_i]$ with

$$n_1 = 6.1 \cdot 10^{22}\,\text{cm}^{-3}, \qquad n_2 = 3.6133 \cdot 10^{21}\,\text{cm}^{-3},$$
$$\mathbf{u}_1 = (9.818 \cdot 10^5,\, 0,\, 0)^\top\, \frac{\text{cm}}{\text{s}}, \qquad \mathbf{u}_2 = (0,\, 0,\, 0)^\top\, \frac{\text{cm}}{\text{s}},$$
$$T_1 = 150\,\text{eV}, \qquad T_2 = 100\,\text{eV}.$$

We simulate this test case using a velocity grid with 48^3 nodes and the second-order IMEX RK scheme from Section 3.3.4 with time step $\Delta t = 0.8\,\text{fs}$.

In Figure 5.12, we plot the evolution of the differences between species temperatures and mean velocities. For constant collision frequencies, the convergence is known to be exponential [CKP20]; this behavior can be clearly observed numerically. However, the convergence of these quantities for the velocity-dependent cross section appears much slower and distinctly different in form. A short exploration of the data suggests that the convergence is neither exponentially fast nor can be described by a power law.

In Figures 5.13 and 5.14, we plot the kinetic distribution of the hydrogen species for $\nu(\mathbf{v})$ and $\hat{\nu}$, the latter giving the slowest relaxation of the velocity-independent collision frequencies. Since the macroscopic quantities of the heavy species (carbon) hardly change, we only show the results for the lighter species (hydrogen).

The relaxation process is weighted by the collision frequencies. Since the velocity-dependent cross section is maximal at $\mathbf{v} = \mathbf{u}_{\text{mix}}$ and decays at larger relative velocity, relaxation to equilibrium in the tails of the distribution is slower when using a velocity-dependent cross section. This can be observed in Figure 5.13 where we show the distributions as functions of $v^{(1)}$ and for fixed $v^{(2)} = u^{(2)} = 0$, $v^{(3)} = u^{(3)} = 0$. And it gets even more evident when comparing Figure 5.13 with Figure 5.14. For the latter, we fix the second and third velocity component at $v^{(2)} = v^{(3)} = -3.58 \cdot 10^7 \frac{\text{cm}}{\text{s}}$, i.e. at the tails for these components.

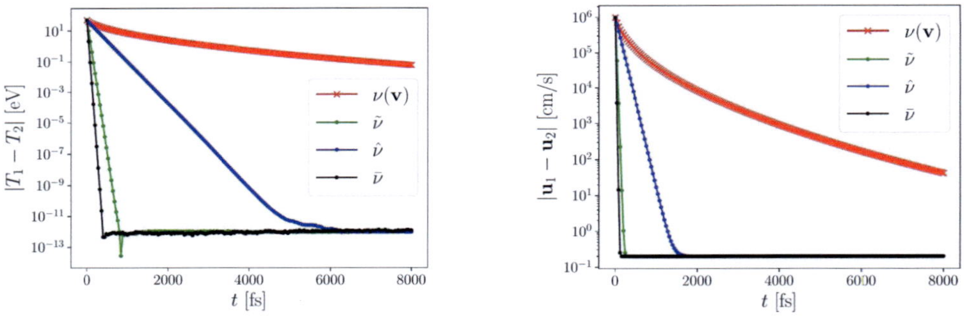

(a) temperature **(b)** mean velocity

Figure 5.12: Evolution of the difference in species temperatures and mean velocities for the Hydrogen-Carbon test case in Section 5.3.2. The convergence for all velocity-independent collision frequencies—$\tilde{\nu}$ in (5.9), $\hat{\nu}$ in (5.11), and $\bar{\nu}$ in (5.12)—appears exponential. However, the convergence for velocity-dependent collision frequency $\nu(\mathbf{v})$ given in (5.8) is significantly longer and notably different.

5.3.3 Sod problem

We run a kinetic version of the well-known Sod problem [Sod78] in the fluid regime (i.e., with large collision frequencies). In the limit of large collision frequencies, the distribution functions can be approximated by Maxwellians:

$$f_i \simeq \mathcal{M}[n_i, \mathbf{u}_i, T_i, m_i].$$

With this approximation, the conservation laws (2.84) reduce to the Euler equations (2.11). We further reduce the problem to the single-species case by assuming $m_1 = m_2 =: m$, $\rho_1 = \rho_2 =: \rho$, $\mathbf{u}_1 = \mathbf{u}_2 =: \mathbf{u}$ and $T_1 = T_2 =: T$. For convenience, we give the Euler equations in this case explicitly. In one space dimension, with 3 translational degrees of freedom, the single-species Euler equations are

$$\partial_t \rho + \nabla_x \cdot (\rho \mathbf{u}) = 0, \tag{5.13a}$$

$$\partial_t (\rho \mathbf{u}) + \nabla_x \cdot (\rho \mathbf{u} \otimes \mathbf{u}) + \nabla_x p = 0, \tag{5.13b}$$

$$\partial_t \left(\frac{\rho |\mathbf{u}|^2}{2} + \frac{3\rho T}{2m} \right) + \nabla_x \cdot \left(\left(\frac{\rho |\mathbf{u}|^2}{2} + \frac{3\rho T}{2m} + p \right) \mathbf{u} \right) = 0, \tag{5.13c}$$

where $p = \frac{\rho T}{m}$ denotes the pressure.

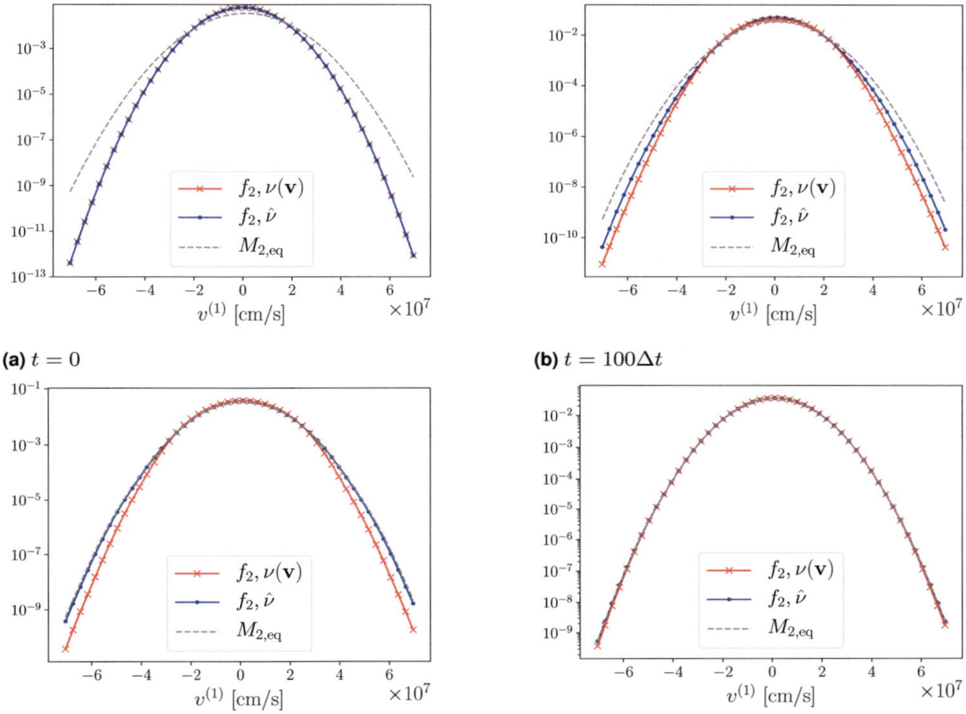

(a) $t = 0$

(b) $t = 100\Delta t$

(c) $t = 500\Delta t$

(d) $t = 10000\Delta t$

Figure 5.13: Relaxation of the kinetic distribution function of hydrogen in the Hydrogen-Carbon test case of Section 5.3.2 with time step $\Delta t = 0.8$ fs. The velocity components $v^{(2)}$ and $v^{(3)}$ are fixed at $u^{(2)} = 0$ and $u^{(3)} = 0$, respectively. The dashed line corresponds to the Maxwellian $M_{2,\mathrm{eq}} = \mathcal{M}[n_2, \mathbf{u}_{\mathrm{mix}}, T_{\mathrm{mix}}, m_2]$, and the blue and red lines are results computed with the velocity-independent collision frequencies $\hat{\nu}$ in (5.11) and the velocity-dependent collision frequencies in $\nu(\mathbf{v})$ (5.8), respectively. The tails of the distribution converge more slowly for the velocity-dependent collision frequencies.

This single-species problem can be implemented with the multi-species model by treating the species identically. We set $m_1 = m_2 = 1$, and the initial data is given by $f_i = \mathcal{M}[n_i, \mathbf{u}_i, T_i, m_i]$ where

$$n_1 = n_2 = 1, \qquad \mathbf{u}_1 = \mathbf{u}_2 = 0, \qquad T_1 = T_2 = 1$$

for $x \leq 0$ and

$$n_1 = n_2 = 0.1, \qquad \mathbf{u}_1 = \mathbf{u}_2 = 0, \qquad T_1 = T_2 = 0.8$$

for $x > 0$.

The equations (5.13) with these initial Riemann data can be solved analytically, see e.g. [LeV02, Tor09]. We provide the analytical solution in Figure 5.15. The two initial states are separated by three intermediate characteristic fields for $t > 0$. From

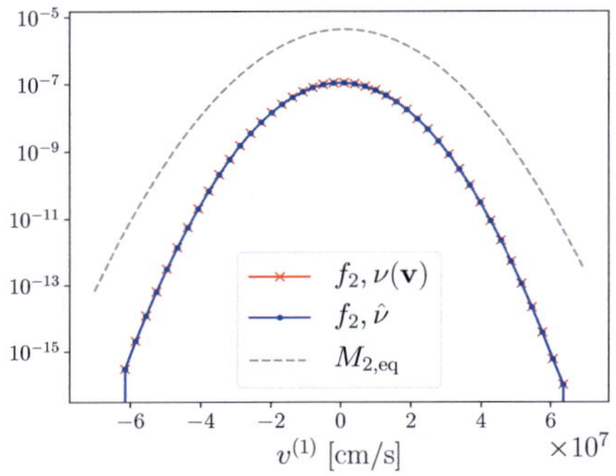

(a) $t = 0$

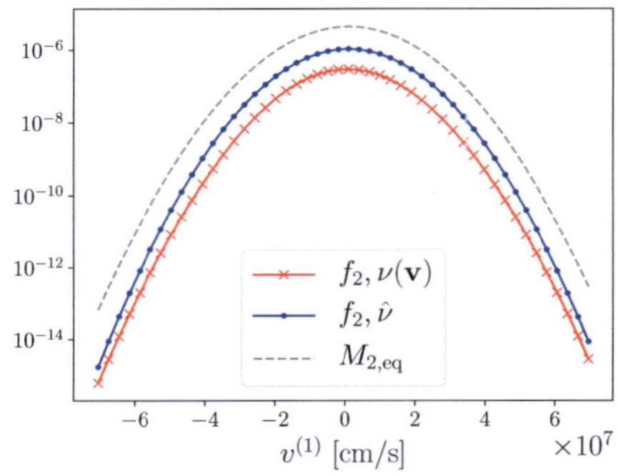

(b) $t = 100\Delta t$

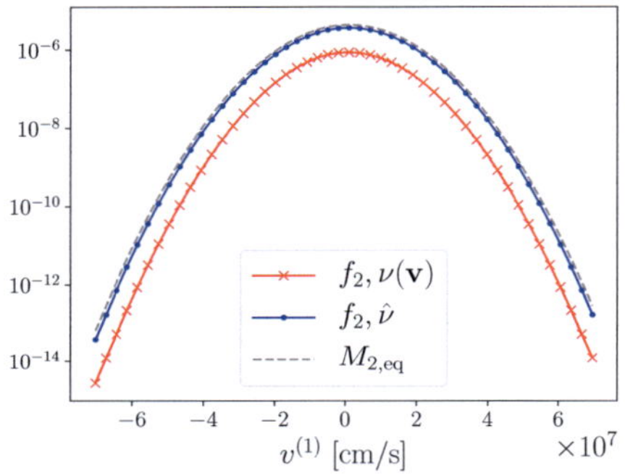

(c) $t = 500\Delta t$

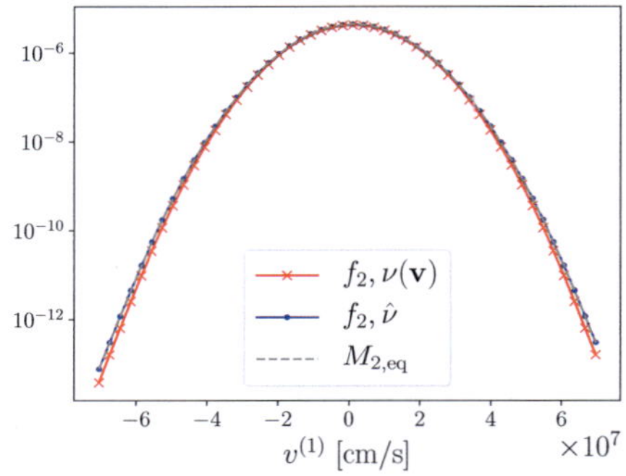

(d) $t = 10000\Delta t$

Figure 5.14: Relaxation of the kinetic distribution function of hydrogen in the Hydrogen-Carbon test case of Section 5.3.2 with time step $\Delta t = 0.8$ fs. The velocity components $v^{(2)}$ and $v^{(3)}$ are fixed at $v^{(2)} = v^{(3)} = -3.58 \cdot 10^7 \, \frac{\text{cm}}{\text{s}}$, respectively, moving away from the center of the distribution function compared to Figure 5.13. The dashed line corresponds to the Maxwellian $M_{2,\text{eq}} = \mathcal{M}[n_2, \mathbf{u}_{\text{mix}}, T_{\text{mix}}, m_2]$, and the blue and red lines are results computed with the velocity-independent collision frequencies $\hat{\nu}$ in (5.11) and the velocity-dependent collision frequencies in $\nu(\mathbf{v})$ (5.8), respectively. This part of the distribution function converges more slowly for the velocity-dependent collision frequencies.

the left to the right, we have the left initial state, a rarefaction wave, a contact discontinuity, and a shock wave separating the right initial state. The analytical solution can be obtained by tracking the characteristics. The wave speeds can be calculated which give us the locations of the fronts. Making use of the Rankine-Hugoniot conditions, the macroscopic quantities can be computed.

We consider two collision frequencies: one that depends on $\mathbf{v}$

$$\nu_{ij}(x, \mathbf{v}, t) = 2 \cdot 10^4 \frac{n_j}{\delta_{ij} + |\mathbf{v} - \mathbf{u}_{\text{mix}}|^3} \tag{5.14}$$

and one that does not

$$\hat{\nu}_{ij}(x, t) = 2 \cdot 10^4 \, \frac{n_j}{\delta_{ij} + \hat{v}^3}, \tag{5.15}$$

where the formula for the averaged relative velocity $\hat{v}$ can be found in (5.10). Again we use the regularization parameter $\delta_{ij} = 0.1 \cdot (\Delta v_{ij})^3$ where $\Delta v_{ij} = \frac{1}{4}\sqrt{T_{\text{mix}}/(2\mu_{ij})}$ and $\mu_{ij} = m_i m_j/(m_i + m_j)$.

The simulations are run using a velocity grid with 48^3 points and 400 equally spaced cells in x. We use the second-order IMEX RK scheme from Section 3.3.4 combined with the second-order FV scheme from Section 3.4, with the limiter given in (3.71). The time step $\Delta t = 9.29 \cdot 10^{-5}$ is set according to the CFL condition in (3.74).

Numerical simulations of the density, mean velocity, and temperature are given in Figure 5.15. We include results using the BGK model with both $\nu(\mathbf{v})$ and $\hat{\nu}$, as well as the analytic solution for the Euler equations (5.13). Both of the collision frequencies $\nu(\mathbf{v})$ and $\hat{\nu}$ give similar results, but the deviations from the Euler solution are more pronounced near the discontinuities in the hydrodynamic model.

For people used to the hydrodynamic regime, the bump in the mean velocity (Figure 5.15 b, d) might be surprising. At the macroscopic level, sharp discontinuities are observed analytically, and there is a constant mean velocity expected across the contact discontinuity. However, at the kinetic level there are too little collisions to push the distribution functions to Maxwellians; discontinuities go with (small) transition regions because particles with e.g. different temperatures mix and do not separate sharply. We see that jumps are smeared out — not only for reasons of numerical dissipation but also due to kinetic theory. As the total energy is conserved, a lower temperature on the RHS of the contact discontinuity goes with a higher kinetic energy and a higher mean velocity, respectively.

5.3.4 Mach 1.7 Shock wave problem

In this example, we compute the flow across a standing Mach 1.7 normal shock wave in a mixture of hydrogen (species 1) and helium (species 2). The shock wave structure is difficult to capture in standard hydrodynamic schemes with a single material/species; in mixtures we further expect species separation to occur due to the mass difference between the two species. The shock conditions are calculated via the Rankine-Hugoniot jump conditions for a monatomic gas [And89]. We take a domain size of 6 microns ($6 \cdot 10^{-4}$ cm) and compute the solution in the frame of the shock. The masses and charge numbers are (units in CGS)

$$m_1 = 1.655 \cdot 10^{-24}\,\text{g}, \quad m_2 = 3.308 \cdot 10^{-24}\,\text{g}, \quad Z_1 = 1, \quad Z_2 = 2.$$

The initial conditions read $f_i = \mathcal{M}[n_i, \mathbf{u}_i, T_i, m_i]$ with

$$n_1 = n_2 = 6.666 \cdot 10^{19}\,\text{cm}^{-3}, \quad u_1 = u_2 = 1.7634411 \cdot 10^7\,\frac{\text{cm}}{\text{s}}, \quad T_1 = T_2 = 100\,\text{eV}$$

for $x \leq 0$ and

$$n_1 = n_2 = 1.308 \cdot 10^{20}\,\text{cm}^{-3}, \quad u_1 = u_2 = 8.985007 \cdot 10^6\,\frac{\text{cm}}{\text{s}}, \quad T_1 = T_2 = 171.32\,\text{eV}$$

for $x > 0$.

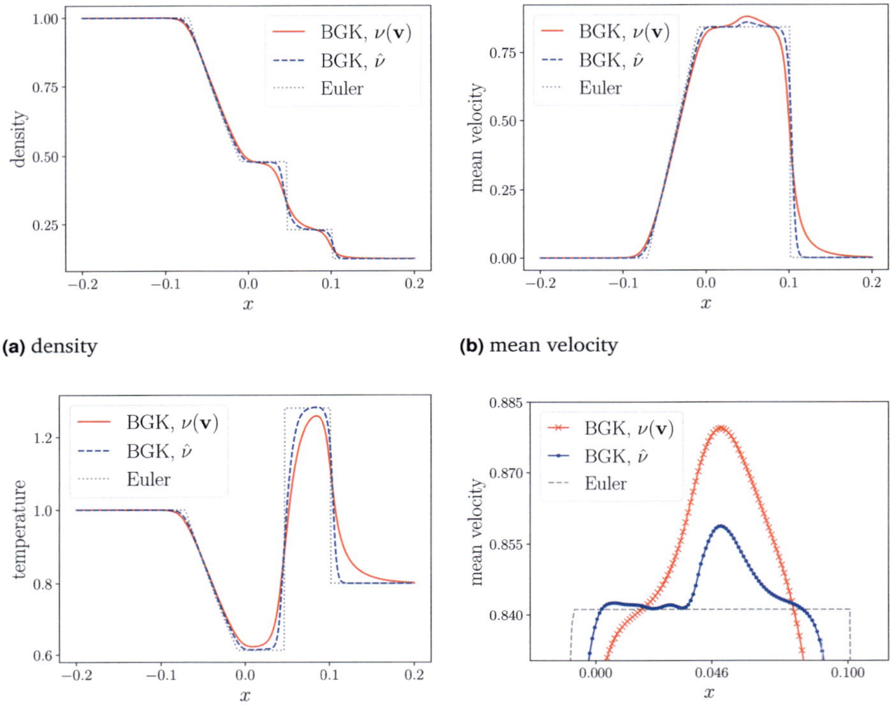

(a) density

(b) mean velocity

(c) temperature

(d) mean velocity, closeup

Figure 5.15: Numerical solution at $t = 0.055$ of the Sod problem in Section 5.3.3. We show results for a 2-species kinetic simulation with using the velocity-dependent collision frequency $\nu(\mathbf{v})$ in (5.14) (red solid line) and the velocity-independent collision frequency $\hat{\nu}$ in (5.15) (dashed blue line). The solutions for both species are identical; we show only the species 1 results. For reference, the exact solution for the Euler equations (5.13) is also provided (dotted gray line). Both kinetic solutions recover the fluid limit fairly well, but the velocity-dependent frequencies contribute to more kinetic behavior around transitions.

The simulations are run using a velocity grid with 48^3 nodes and a spatial mesh with 200 cells. We use the second-order IMEX RK scheme from Section 3.3.4 and the second-order spatial discretization in Section 3.4, with the limiter given in (3.71). The time step $\Delta t = 22$ fs is set according to the CFL condition in (3.74).

In Figure 5.16 we compare numerical results at time $t = 5.390$ ps using the velocity-dependent collision frequency $\nu(\mathbf{v})$, given in (5.8), with those using the constant collision frequencies $\hat{\nu}$, given in (5.11). In addition to these results, we plot the relative difference

$$r(q) = \frac{q(\{\hat{\nu}_{ij}\}) - q(\{\nu_{ij}(\mathbf{v})\})}{|q(\{\hat{\nu}_{ij}\})| + |q(\{\nu_{ij}(\mathbf{v})\})|} \tag{5.16}$$

for the densities ($q = n_i$), mean velocities ($q = u_i^{(1)}$), and temperatures ($q = T_i$). As expected, both the velocity-dependent and constant collision frequency models show a species separation. For all hydrodynamic quantities, the differences are within a few percent. While we expect a difference in output profiles between the two models due to the tail particles relaxing more slowly than the bulk, it is likely that the collision frequencies outside of the 'kinetic' region of the shock interface are high enough to suppress large deviations from equilibrium.

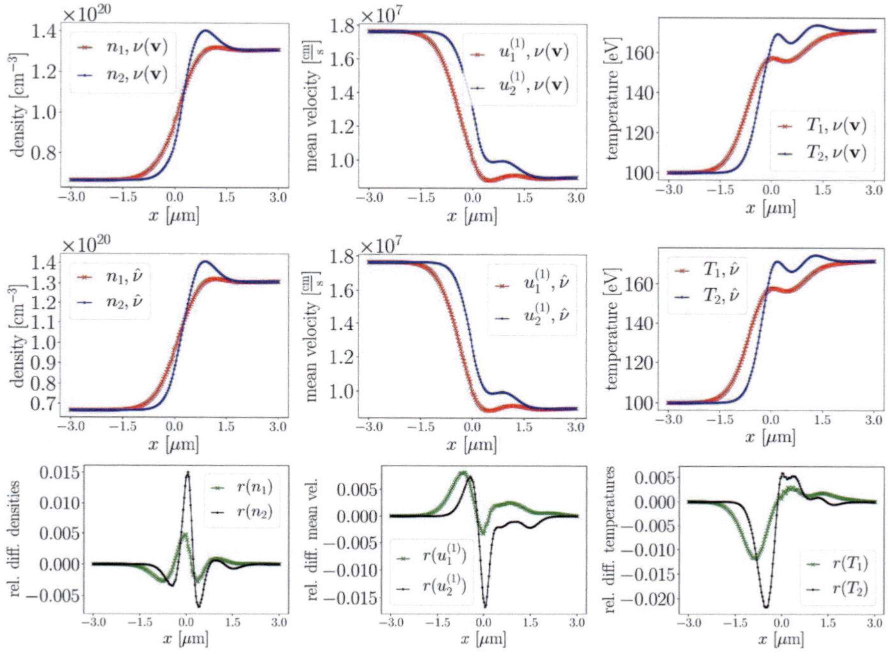

Figure 5.16: Fluid quantities at time $t = 5.39$ ps for the Mach 1.7 shock wave problem from Section 5.3.4. The initial Riemann problem forms into a standing shock wave, where hydrogen and helium separate from each other near the interface. Top row: velocity-dependent collision frequencies $\nu_{ij}(\mathbf{v})$, given in (5.8); middle row: constant collision frequencies $\hat{\nu}_{ij}$, given in (5.11); bottom row: relative differences (see (5.16)). These differences are typically within 4%.

5.3.5 Mach 4 Shock wave problem

We repeat the shock wave problem from the previous case but increase the shock speed in the mixture to Mach 4, with the expectation that the distributions will be further out of equilibrium than in the previous case. The masses and charges of the species are the same as in the Mach 1.7 case, but we widen the domain size to 12 microns and modify the initial conditions to construct a Mach 4 shock, again using the Rankine-Hugoniot relations. Specifically $f_i = \mathcal{M}[n_i, \mathbf{u}_i, T_i, m_i]$ where

$$n_1 = n_2 = 3.3488 \cdot 10^{19}\,\text{cm}^{-3}, \qquad u_1 = u_2 = 5.06 \cdot 10^7\,\frac{\text{cm}}{\text{s}}, \qquad T_1 = T_2 = 100\,\text{eV},$$

for $x \leq 0$ and

$$n_1 = n_2 = 1.128 \cdot 10^{20}\,\text{cm}^{-3}, \qquad u_1 = u_2 = 1.50 \cdot 10^7\,\frac{\text{cm}}{\text{s}}, \qquad T_1 = T_2 = 586.3\,\text{eV}.$$

and for $x > 0$.

As in the previous case, simulations are run using a velocity grid with 48^3 nodes and a spatial mesh with 200 cells. We use the second-order IMEX RK scheme from Section 3.3.4 and the second-order spatial discretization in Section 3.4, with the limiter given in (3.71). The time step $\Delta t = 25$ fs is set according to the CFL condition in (3.74).

In Figure 5.17 we compare numerical results at time $t = 6.345$ ps using the velocity-dependent collision frequency $\nu(\mathbf{v})$, given in (5.8), with those using the constant collision frequencies $\hat{\nu}$, given in (5.11). We again observe the evolution towards a standing shock wave for both the velocity-dependent collision frequency $\nu(\mathbf{v})$ and the constant collision frequency $\hat{\nu}$. As in the Mach 1.7 case above, while we expect a difference in output profiles between the two models due to the tail particles relaxing more slowly than the bulk, it is likely that the collision frequencies outside of the 'kinetic' region of the shock interface are high enough to suppress large deviations from equilibrium for this test problem.

The shock wave problems are set up using hydrodynamic theory. Hence, the fluids are already close to equilibrium suppressing the effects of the velocity-dependent collision frequencies. The following test cases, in contrast, illustrate that constant collision frequencies may lose details from the kinetic level which are visible with velocity-dependent ones.

5.3.6 Interpenetration problem: high density

Standard hydrodynamic models have great difficulty in capturing interpenetrating flows of rarefied gases. For example in ICF simulations, colliding streams of blown-off hohlraum wall particles and capsule ablator particles result in an unphysical density spike due to the lack of interpenetration in hydrodynamic models, which interferes with laser energy propagation in the integrated simulation. This discrepancy has been proposed as a cause of symmetry discrepancies in capsule drive between experiments and simulations in ICF [BHLPD+15].

For this numerical example, we simulate the dynamics of two counter-streaming beams of different species. We take a domain size of 50 microns ($50 \cdot 10^{-4}$ cm) and compute the solution when hydrogen (species 1) interpenetrate with helium (species 2) particles. We include a trace amount of each species in the whole domain as a background for ease of computation. The masses and charge numbers are (units in CGS)

$$m_1 = 1.655 \cdot 10^{-24}\,\text{g}, \quad m_2 = 3.308 \cdot 10^{-24}\,\text{g}, \quad Z_1 = 1, \quad Z_2 = 2.$$

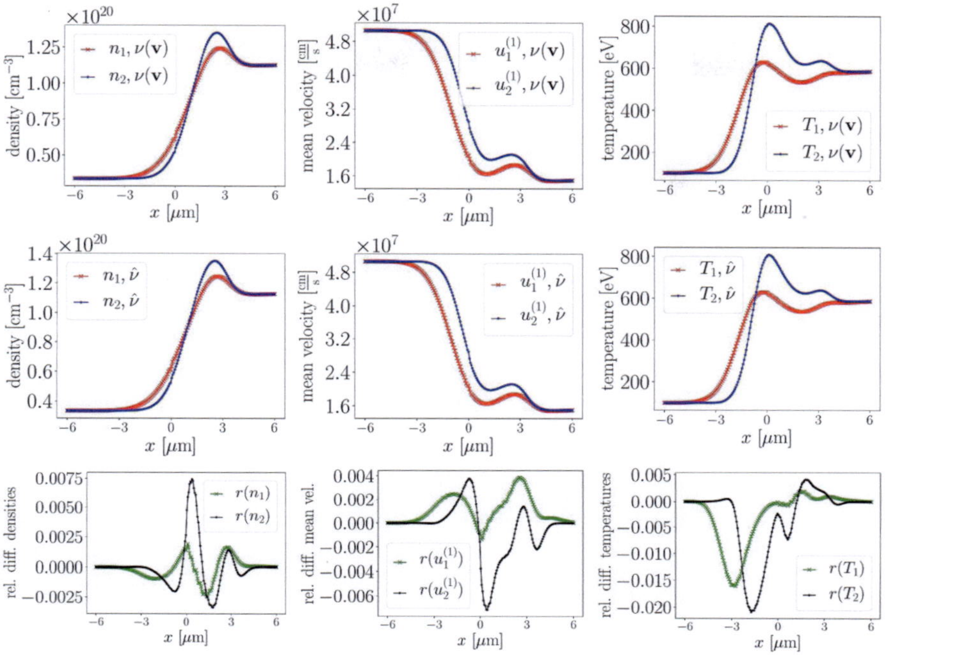

Figure 5.17: The fluid quantities for the Mach 4 shock wave problem from Section 5.3.5 (Mach 4) are presented at time $t = 6.345$ ps. The initial Riemann problem forms into a standing shock wave. Top row: velocity-dependent collision frequencies $\nu_{ij}(\mathbf{v})$, given in (5.8); middle row: constant collision frequencies $\hat{\nu}_{ij}$, given in (5.11); bottom row: relative differences (see (5.16)). The results for the velocity-dependent collision frequency and the constant collision frequency $\hat{\nu}_{ij}$ look very similar at first glance. However, the relative differences clarify the disparities. They differ up to 4 %.

The initial conditions are $f_i = \mathcal{M}[n_i, \mathbf{u}_i, T_i, m_i]$ with

$$n_1 = 10^{20}\,\text{cm}^{-3}, \quad n_2 = 10^{17}\,\text{cm}^{-3}, \quad u_1 = u_2 = 2.2 \cdot 10^6\,\frac{\text{cm}}{\text{s}}, \quad T_1 = T_2 = 10\,\text{eV}$$

for $x \leq 0$ and

$$n_1 = 10^{17}\,\text{cm}^{-3}, \quad n_2 = 10^{20}\,\text{cm}^{-3}, \quad u_1 = u_2 = -2.2 \cdot 10^6\,\frac{\text{cm}}{\text{s}}, \quad T_1 = T_2 = 10\,\text{eV}$$

for $x > 0$.

The simulations are run using a velocity grid with 48^3 nodes and a spatial mesh with 200 cells. We use the second-order IMEX RK scheme from Section 3.3.4 and the second-order spatial discretization in Section 3.4, with the limiter given in (3.71). The time step $\Delta t = 806$ fs is set according to the CFL condition in (3.74).

In Figure 5.18, we compare the numerical results at time $t = 120.870$ ps using the velocity-dependent collision frequency $\nu(\mathbf{v})$, given in (5.8), with those using the constant collision frequencies $\hat{\nu}$, given in (5.11). The lighter hydrogen species shows

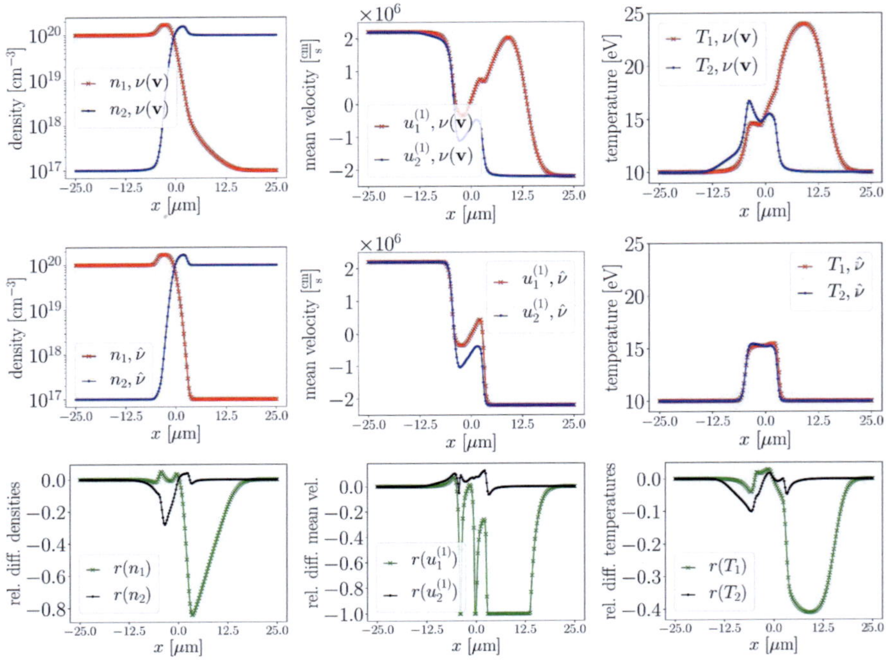

Figure 5.18: The fluid quantities for the interpenetration problem from Section 5.3.6 are presented at time $t = 120.870$ ps. First row: velocity-dependent collision frequencies $\nu_{ij}(\mathbf{v})$, given in (5.8). Second row: constant collision frequencies $\hat{\nu}_{ij}$, given in (5.11). Third row: relative differences between rows 1 and 2 according to (5.16). Red line: hydrogen. Blue line: helium. Variations in the collision frequency induce significant differences in the profile of the hydrogen, which penetrates much further into the right side of the domain when the collision frequency is velocity-dependent. Due to relatively higher mass and charge state, the helium species undergoes more collisions and is less sensitive to variations in the collision frequency.

a significant difference in profiles between the two species, and displays much more penetration into the helium beam. Due to its relatively higher mass and charge state, the helium species is much more collisional than the hydrogen species, and presents a more hydrodynamic-like profile.

To highlight the different behavior due to the respective collision frequency, we provide a direct comparison of the simulations with the velocity-dependent collision frequency and with the constant collision frequency in Figure 5.19. It is clearly visible that the hydrogen species behaves significantly different.

5.3.7 Interpenetration problem: low density

We repeat the interpenetration problem from above but reduce the initial densities by two orders of magnitude, which leads to fewer collisions. We expect to see a greater interpenetration of the two beams with less of a density spike at the interface point.

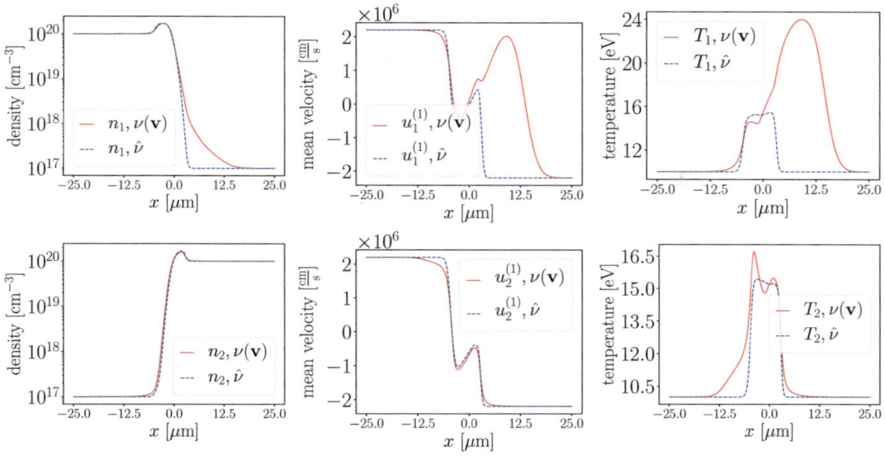

Figure 5.19: Comparison of the fluid quantities for the interpenetration problem from Section 5.3.6 when run with velocity-dependent collision frequencies $\nu_{ij}(\mathbf{v})$, given in (5.8), and constant collision frequencies $\hat{\nu}_{ij}$, given in (5.11). The hydrogen species behaves significantly different.

The domain size, masses and charges are the same as before. The initial conditions are $f_i = \mathcal{M}[n_i, \mathbf{u}_i, T_i, m_i]$ with

$$n_1 = 10^{18}\,\text{cm}^{-3}, \quad n_2 = 10^{15}\,\text{cm}^{-3}, \quad u_1 = u_2 = 2.2 \cdot 10^6\,\frac{\text{cm}}{\text{s}}, \quad T_1 = T_2 = 10\,\text{eV}$$

for $x \leq 0$ and

$$n_1 = 10^{15}\,\text{cm}^{-3}, \quad n_2 = 10^{18}\,\text{cm}^{-3}, \quad u_1 = u_2 = -2.2 \cdot 10^6\,\frac{\text{cm}}{\text{s}}, \quad T_1 = T_2 = 10\,\text{eV}$$

for $x > 0$.

As before, the simulations are run using a velocity grid with 48^3 nodes and a spatial mesh with 200 cells. We use the second-order IMEX RK scheme from Section 3.3.4 and the second-order spatial discretization in Section 3.4, with the limiter given in (3.71). The time step $\Delta t = 806$ fs is set according to the CFL condition in (3.74).

In Figure 5.20, we compare the numerical results at time $t = 120.870$ ps using the velocity-dependent collision frequency $\nu(\mathbf{v})$, given in (5.8), with those using the constant collision frequencies $\hat{\nu}$, given in (5.11). As expected, we see more interpenetration than in the high density test case. As in the higher density test case, we see more significant differences in the lighter species of the mixture; the hydrogen species penetrates more into the right side of the domain when the collision frequency is velocity-dependent. Due to relatively higher mass and charge state, the helium species is more collisional. Furthermore, the density spike at the interface seen in the high density case has mostly disappeared.

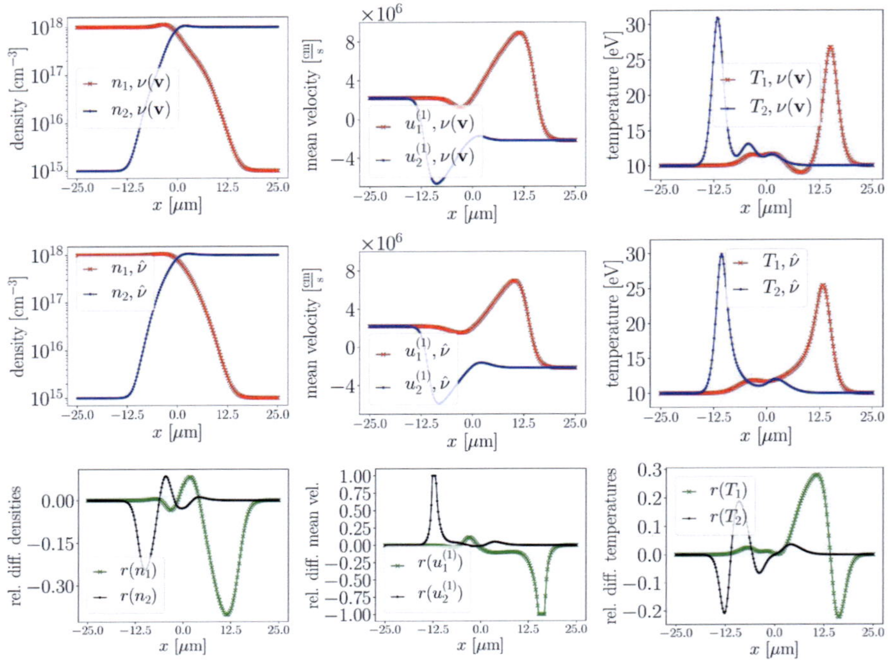

Figure 5.20: The fluid quantities for the lower density interpenetration problem from Section 5.3.7 are presented at time $t = 120.870$ ps. First row: velocity-dependent collision frequencies $\nu_{ij}(\mathbf{v})$, given in (5.8). Second row: constant collision frequencies $\hat{\nu}_{ij}$, given in (5.11);. Third row: relative differences between rows 1 and 2 according to (5.16). Red line: hydrogen. Blue line: helium. As expected, we see more interpenetration than in the high density test case. However, the relative sensitivity of hydrogen to the velocity-dependent collision frequency is less dramatic.

To emphasize the different behavior due to the respective collision frequency, we provide a direct comparison of the simulations with the velocity-dependent collision frequency and with the constant collision frequency in Figure 5.21.

We have provided several test cases for the model with velocity-dependent collision frequency from Section 2.2. Now, we study numerical set-ups for the quantum model.

5.4 Numerical results for the quantum multi-species BGK model

We run simulations for the quantum multi-species BGK model from Section 2.3. We only focus on space homogeneous tests. In Section 5.4.1, we validate the results by comparing them with the analytic decay rates (2.142) and (2.145). In Section 5.4.2, we rerun the test case from Section 5.2.3 and encounter different types of species.

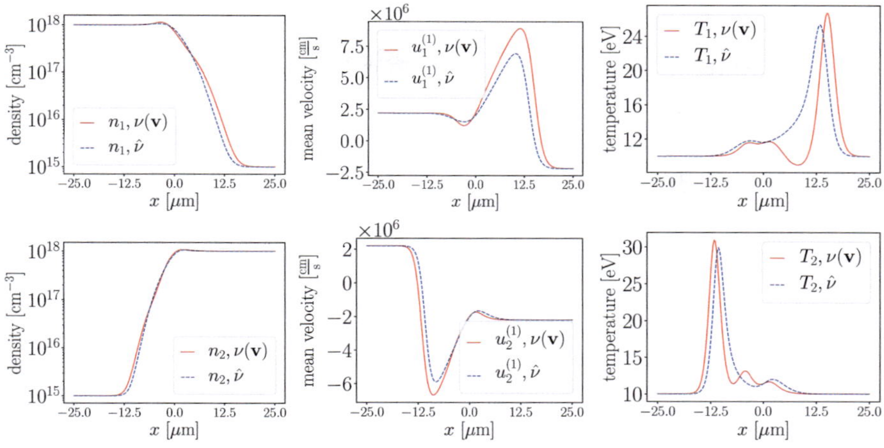

Figure 5.21: Comparison of the fluid quantities for the interpenetration problem from Section 5.3.7 when run with velocity-dependent collision frequencies $\nu_{ij}(\mathbf{v})$, given in (5.8), and constant collision frequencies $\hat{\nu}_{ij}$, given in (5.11). The hydrogen species still behaves different but less dramatic compared to the high density test case.

5.4.1 Quantum decay rates

In this test case, we illustrate the correct behavior of our numerical scheme for quantum particles. Moreover, we verify the decay rates which are given analytically in Section 2.3.2.

Initially, we set the distribution functions to Maxwellians $f_i = \mathcal{M}[n_i, \mathbf{u}_i, T_i, m_i]$ with

$$m_1 = 1.0, \quad n_1 = 1.0, \quad \mathbf{u}_1 = (0.5, 0, 0)^\top, \quad T_1 = 1.0,$$
$$m_2 = 1.5, \quad n_2 = 1.2, \quad \mathbf{u}_2 = (0.1, 0, 0)^\top, \quad T_2 = 0.5,$$

inspired by the test case in Secton 5.2.1. By choosing these initial data, we do not incorporate further physical details (e.g. for a specific quantum regime), but we want to illustrate the basic properties of the model and scheme, respectively.

For the simulation, we use a momentum grid with 48^3 nodes and the first-order splitting scheme from Section 3.3.1 with the time step $\Delta t = 0.01$.

We run the simulation for any combination of classical particles, fermions and bosons. In Figure 5.22, we illustrate the evolution of the entropy and the entropy dissipation exemplary for the interactions of fermions with fermions. Additionally, we demonstrate the conservation properties in Figure 5.23 where the numerical oscillations in mass, total momentum and total energy are only of the order 10^{-14}.

In Figure 5.24, we show the behavior of the mean velocities converging exponentially fast to a common value. The numerical decay rate matches the analytical one (2.142) very well. Since the decay rate is independent of the type of the species, we only display the rate for the interactions of fermions with fermions.

Figure 5.25 illustrates the behavior of the temperatures where we distinguish between the kinetic temperatures T_i and the physical temperatures ϑ_i, see Remark 2.3.10. In the first column, we observe a gap between the final values of the species' kinetic temperatures whenever a quantum particle is involved. This is also visible in the middle column where the decay rate for the kinetic temperatures is illustrated. Numerical and analytical results coincide very well, and the difference converges to a constant value for quantum particles. Such behavior of the kinetic temperatures for quantum particles comes by an additional term for the decay rates (2.145) which vanishes for classical-classical interactions, see Remark 2.3.23. We compare the results for the kinetic temperatures with the physical temperatures in the last column. Even though the kinetic temperatures behave differently for quantum particles, the physical temperatures converge to a common value in all cases.

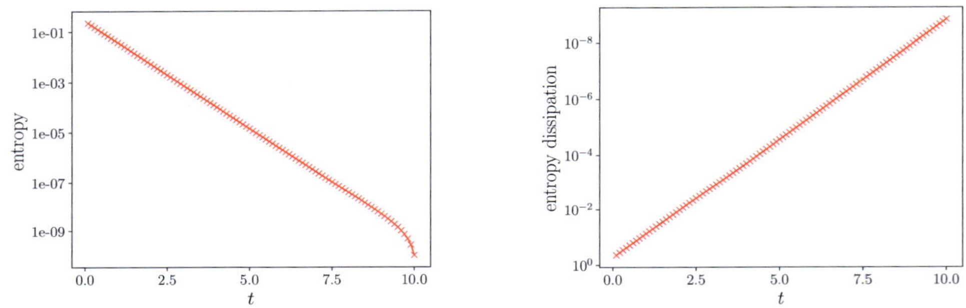

(a) entropy **(b)** entropy dissipation

Figure 5.22: Entropy and entropy dissipation for the test case in Section 5.4.1, exemplary for fermion-fermion interactions. As predicted by the theory, the entropy decays monotonically.

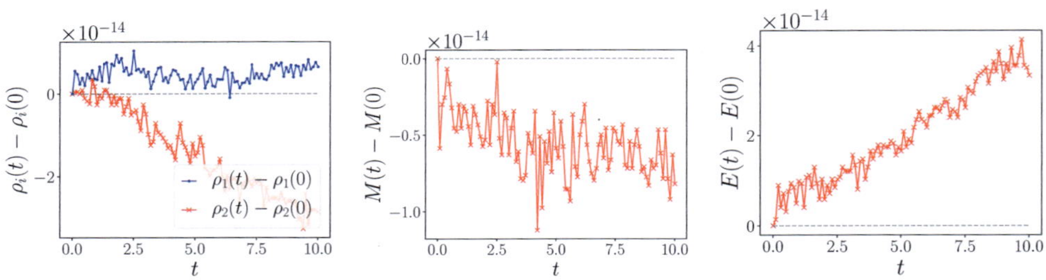

(a) species mass **(b)** total momentum **(c)** total energy

Figure 5.23: Global conservation properties for the test case in Section 5.4.1. The mass densities of each species, the total momentum (M) and total energy (E) have small oscillations of the order of 10^{-14}.

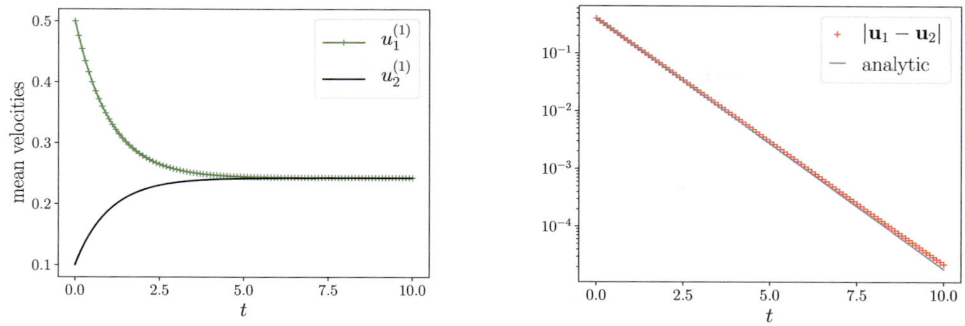

Figure 5.24: Mean velocities for the test case in Section 5.4.1, exemplary for fermion-fermion interactions as the convergence is independent of the type of species. The mean velocities converge exponentially fast to a common value, and the numerical decay rate hits the analytical one very well.

5.4.2 Sulfur-Fluorine-electrons quantum test case

We repeat the test case from Section 5.2.3, but we treat the electrons like fermions; the ions are kept to be classical particles. Moreover, we use constant collision frequencies $\nu_{ij} = 1$ because of the underlying model from Section 2.3.2.

The masses and charge numbers of the species are

$$\begin{aligned} &m_S = 32.07\text{u} - 11m_e, \quad m_F = 19\text{u} - 7m_e, \quad m_e = m_e, \\ &Z_S = 11, \quad Z_F = 7, \quad Z_e = -1 \end{aligned}$$

with the atomic mass $\text{u} = 1.6605 \cdot 10^{-24}$ g. Initially, we assume Maxwellian distributions $f_i = \mathcal{M}[n_i, \mathbf{u}_i, T_i, m_i]$

$$\begin{aligned} &n_S = 10^{19}\,\text{cm}^{-3}, \quad n_F = 6 \cdot 10^{19}\,\text{cm}^{-3}, \quad n_e = 53 \cdot 10^{19}\,\text{cm}^{-3}, \\ &\mathbf{u}_S = \mathbf{u}_F = \mathbf{u}_e = 0\,\frac{\text{cm}}{\text{s}}, \\ &T_S = T_F = 15\,\text{eV}, \quad T_e = 100\,\text{eV}. \end{aligned}$$

We use momentum grids with 48^3 nodes for each species, and we use the second-order IMEX RK scheme from Section 3.3.4 with time step $\Delta t = 0.01$ fs.

The species are initialized with vanishing mean velocity, and the mean velocities stay zero. We give the evolution of the temperatures in Figure 5.26. The kinetic temperatures approach the mixture temperature (2.9), but they do not converge to a common value which matches the experiences from the previous test case (see Figure 5.26 (a),(b)). The relaxation process is much faster than in Section 5.2.3. This is not surprising because the collision frequencies $\nu_{ij} = 1$ are much larger than before. Additionally, we observe that the sulfur and fluorine particles relax in a slightly different way (see Figure 5.26 (c)), whereas they always hit the same values in the purely classic test case in Section 5.2.3.

Hence, we see an impact when treating electrons like fermions. This motivates for future studies.

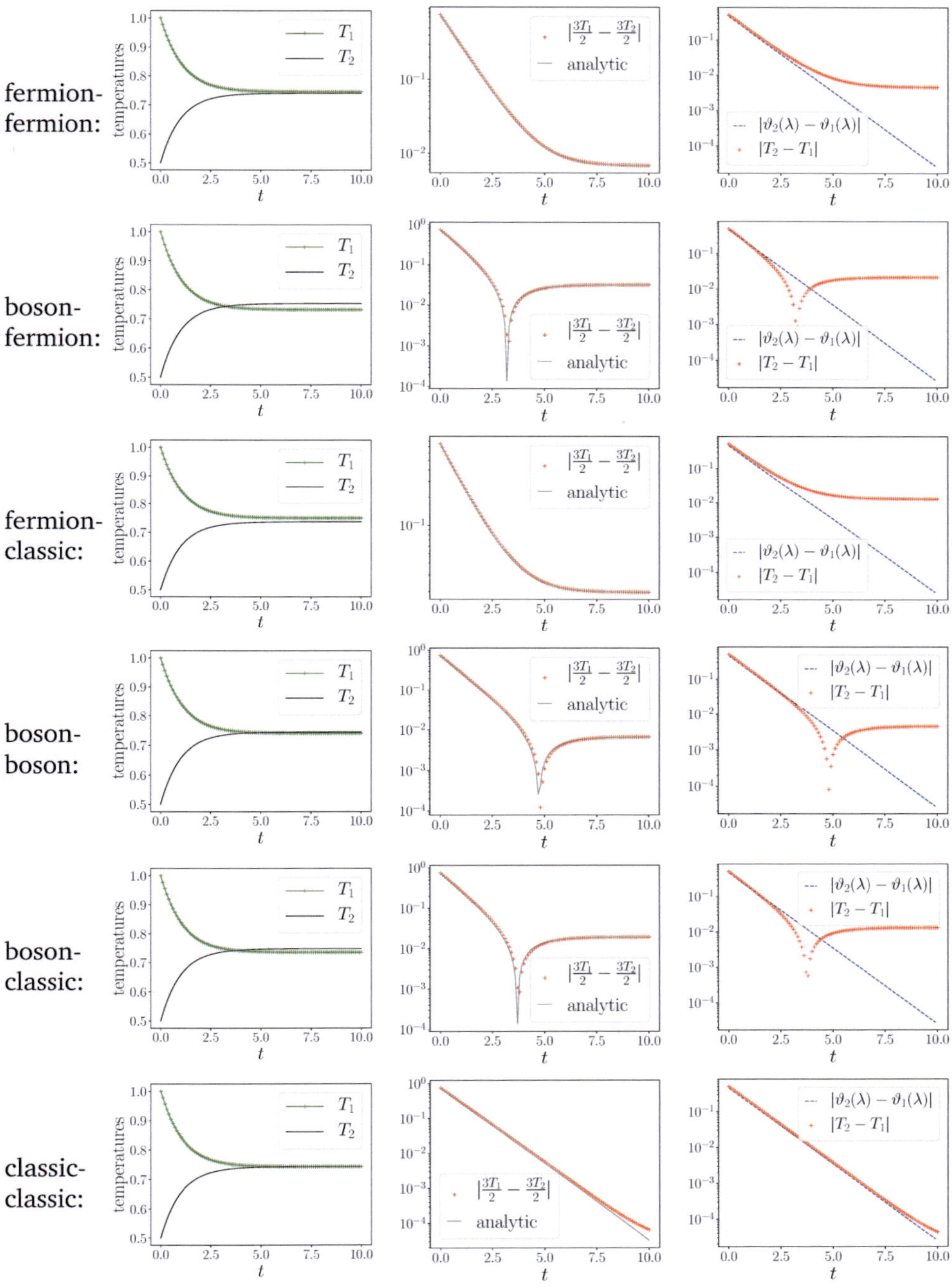

Figure 5.25: Temperatures for the test case in Section 5.4.1. First column: kinetic temperatures; there remains a gap between the final values whenever a quantum particle is involved. Second column: decay rates for kinetic temperatures in logarithmic scale; numerical and analytical values coincide very well. Third column: comparison of kinetic and physical temperature in logarithmic scale; the physical temperatures ϑ converge exponentially fast, whereas the kinetic temperatures T behave differently for quantum particles.

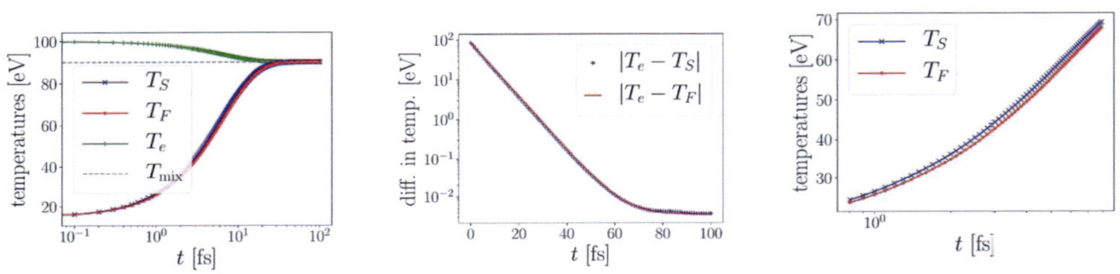

(a) kinetic temperatures

(b) decay rate

(c) kinetic temperatures, closeup

Figure 5.26: Evolution of the kinetic temperatures for the Sulfur-Fluorine-electrons quantum test case in Section 5.4.2 using the constant collision frequencies $\nu_{ij} = 1$ and treating the electrons like fermions. It seems as if the kinetic temperatures hit the mixture temperature (2.9) as in the classical test case (a). However, we observe that the differences in the kinetic temperatures only decay exponentially fast until they stay constant (b). Additionally, the relaxation process of sulfur particles differs from the relaxation process of fluorine particles (c).

Chapter 6

Conclusions and Outlook

We have given an introduction into kinetic theory with a focus on BGK-type equations. A consistent multi-species BGK model with velocity-dependent collision frequency is described in more detail including a motivation how to define collision frequencies for Coulomb interactions in plasmas. Furthermore, we consider a quantum multi-species BGK model and provide analytical decay rates for the mean velocities and kinetic temperatures.

We have developed a numerical scheme which can be applied to both of the above models. As the target functions depend only implicitly on the distribution functions, a new approach for the implication of the target functions is required. We provide a general implicit solver which determines the target functions via a convex minimization problem. This procedure mimics the dual of the minimization problem which defines the theoretical model and also guarantees the conservation of mass, total momentum and total energy at the discrete level during the relaxation process. Using a standard FV method ensures the conservation properties during the transport process. The conservation properties at the discrete level are proven analytically. The positivity of distribution functions holds rigorously coupled to a possible time step restriction. We prove that a discrete entropy dissipation property is fulfilled for a first-order scheme. Second-order schemes are used for improved accuracy.

We have performed several numerical tests in order to illustrate the properties of the BGK models and our numerical scheme. At first, we run test cases for a multi-species BGK model with constant collision frequency comparing the results with analytical values or simulations from the literature. The results are in good agreement.

The main part is given by several tests for the multi-species BGK model with velocity-dependent collision frequencies where we use collision frequencies that are suitable for plasmas and characterized by a slower relaxation in the tails of the distribution functions. We compare the results with simulations using velocity-independent collision frequencies of comparable size. For space homogeneous setups, we illustrate that the velocity-dependent collision frequencies induce a slower relaxation to equilibrium in the tails of the kinetic distributions; additionally, the convergence of the mean velocities and temperatures is slower and different in form than for constant collision frequencies. We examine several Riemann problems. The standard Sod shock tube problem is performed: the general fluid shock structure is recovered, however kinetic effects are better observable for velocity-dependent collision frequencies. Moreover, we run variations to the Sod problem involving mixtures. The profiles for the Mach shock wave problems show close agreement between simulations using velocity-dependent and velocity-independent collision frequencies. Nevertheless, for the interpenetration problem the effect of velocity-dependent collision frequencies on the lighter species (in mass and charge state) are meaningful.

Furthermore, we illustrate the evolution of mean velocities and temperatures in a space homogeneous setting when quantum particles are involved. The behavior of the mean velocities is unaffected, and they converge exponentially fast to a common value like in the classical case — in contrast to the temperatures. We need to distinguish between kinetic and physical temperatures. While the physical ones do converge to a common value, the kinetic ones behave differently resulting in a gap between the species' kinetic temperatures in equilibrium. The different behavior of a relaxation process of purely classical particles compared to a relaxation process of classical particles and fermions inspires to further studies.

The BGK approximation of the Boltzmann operator is appreciated among other things for the ability of efficient simulations. However, extending the models by allowing for velocity-dependent collision frequencies or including quantum effects is not without additional costs. For our proposed scheme, the implication of the interaction operator is realized by a minimization problem which requires a Newton solver. For the gradient and Hessian of the potential function, integrals in velocity space need to be calculated via a quadrature. A more efficient implementation of the optimization algorithm [SBT17, AHOT14, AHT12, KHH15, Abr07] including a more advanced handling of the velocity grid and quadrature rules is necessary in order to accelerate the solution procedure. Even though the extensions go with additional numerical costs, the models still profit from better scaling properties than the original Boltzmann equations. We confirm this statement in the following. For evaluating the Boltzmann collision operator, the fastest algorithms are spectral methods whose complexity for general collision kernels scale like $\mathcal{O}(MN^4 \log N)$. This can be reduced to $\mathcal{O}(MN^3 \log N)$ for specialized kernels [GHHH17, GT09, PR00, MP06]. Here, N is the number of points in each dimension of the velocity grid, and M is the number of quadrature points for the approximation of the integrals over the unit sphere $\mathbb{S}^2$ in $\mathbb{R}^3$. The size of M is problem dependent, but typically it is $N \leq M \ll N^2$ [MP06]. In comparison, the number of operations needed for evaluating the BGK operator in our scheme scales like $\mathcal{O}(N^3)$. Thus, our presented method is more expensive than for standard BGK models but still of lower computational complexity than for the Boltzmann operator. Moreover, grid resolution for multi-species Boltzmann collision operators introduce expensive requirements for problems with significant differences in species masses [MTH$^+$14]. Whereas we keep the advantage of BGK models that the equations can be discretized on separate velocity grids.

Another possible feature of BGK models is the AP property. In this thesis, we have not examined this attribute, but it would be worthwhile for future works, also in context of the efficiency.

The idea of the implicit solver is not limited to a specific order of the underlying scheme nor to a specific discretization technique in time. Hence, the use of high-order schemes in time is straight-forward. The extension to higher order in space is more difficult because the relaxation operator also needs to undergo a high-order reconstruction. Alternatively, one might use an SL approach.

The enlarged class of possible collision frequencies as well as including quantum effects are physically motivated. Both extensions represent attractive options for exploring more phenomena in the kinetic regime. Especially the quantum model asks for further tests with more elaborated initial data and in more general settings (e.g. shock waves).

Additionally, the models can be improved. A plasma consists of charged particles which required to include a force term with an electric (and magnetic) field. The associated transport in velocity space can be easily incorporated in the presented numerical method. We have started a short excursion in Appendix A. We provide initial tests where we simulate the phenomenon of Landau damping and consider Landau damping coupled with relaxation. The numerical results match our expectations and form a basis for further simulations. For future studies, it would be interesting to study different species (in charge and mass) in more general set-ups and to use more advanced methods for the transport terms.

Appendices

A On damping of the electric field

In our scheme, we have not considered possible force terms, yet. Especially for charged particles, this is a severe simplification. When a force term is encountered, an advection in velocity space takes place. This is described by the additional term

$$\frac{\mathbf{F}_i}{m_i} \cdot \nabla_{\mathbf{v}} f_i$$

on the LHS of the kinetic equation, see Sections 2.1.1 and 2.1.2.

We now include a force term by the coupling to a self-consistent electric field which has been described in Section 2.1.2. For simplicity, we only consider classical particles. We implement a simplified version of the Lorentz force (2.17), i.e.

$$\mathbf{F}_i = Z_i e \mathbf{E}$$

where Z_i denotes the charge state of species i, e the elementary charge and $\mathbf{E}$ the (self-consistent) electric field. In our one-dimensional setting in space, the governing equations for the electric field[1] reduce to (2.24), i.e.

$$\partial_x E(x,t) = \frac{q(x,t)}{\varepsilon_0} \tag{6.1}$$

with the charge density $q = \sum_i Z_i e n_i$ and the vacuum permittivity ε_0. We assume periodic boundary conditions and the zero-mean electrostatic condition

$$\int E \,\mathrm{d}x = 0. \tag{6.2}$$

Since the force $F(x,t)$ acts only in the x-direction, the advection occurs only in the first component of $\mathbf{v}$: $v^{(1)}$. For ease in notation, we write $v = v^{(1)}$.

The system of PDEs (3.29) is extended to

$$\begin{aligned} \partial_t f_1 + \mathcal{T}(f_1) &= \mathcal{R}_1(f_1, f_2) \\ \partial_t f_2 + \mathcal{T}(f_2) &= \mathcal{R}_2(f_2, f_1) \end{aligned} \tag{6.3}$$

with the transport operator

$$\mathcal{T}(f_i) = v\,\partial_x f_i + \frac{F_i}{m_i}\partial_v f_i$$

1 In this section, E stands for the electric energy which is *not* the energy density of a species.

and the relaxation operator

$$\mathcal{R}_i(f_i, f_j) = \nu_{ii}(\mathcal{A}_{ii} - f_i) + \nu_{ij}(\mathcal{A}_{ij} - f_i).$$

The system (6.3) coupled to (6.1) is also called Vlasov-Poisson-BGK equations. For references concerning numerics, see e.g. [HHM17b, Ful21]. We apply the discretization techniques which have been presented in Sections 3.2–3.5. The remnant parts which come by the force term are discussed in the following.

A.1 Numerical fluxes

We discretize the force term by a FV method. We interpret the values at the discrete-velocity grid as cell-averaged quantities, which is second-order accurate. The advection in v is discretized by numerical fluxes. Applying the same approach used for the advection in x to the advection in v, we obtain

$$\frac{F_i}{m_i}\partial_v f \approx \frac{1}{\Delta v_i}\left(\mathscr{F}_{i,q_1+\frac{1}{2}} - \mathscr{F}_{i,q_1-\frac{1}{2}}\right)$$

where

$$\mathscr{F}_{i,q_1+\frac{1}{2}}(g) = \frac{F_i}{2m_i}\left(g_{q_1+1} + g_{q_1}\right) - \frac{|F_i|}{2m_i}\left(g_{q_1+1} - g_{q_1} - \phi_{q_1+\frac{1}{2}}(g)\right)$$

and the flux limiter ϕ is given in (3.71).

A simple Forward Euler update (3.2) of

$$\partial_t f + \frac{F_i}{m_i}\partial_v f = 0$$

reads

$$f_{i,q_1}^{\ell+1} = f_{i,q_1}^{\ell} - \frac{\Delta t}{\Delta v_i}\left(\mathscr{F}_{i,q_1+\frac{1}{2}}(f_i^{\ell}) - \mathscr{F}_{i,q_1-\frac{1}{2}}(f_i^{\ell})\right). \tag{6.4}$$

The positivity of f_i is guaranteed in (6.4) by enforcing the CFL condition

$$\Delta t < \alpha \min_i \left\{ \frac{m_i \Delta v_i}{\max_x |F_i|} \right\} \tag{6.5}$$

with $\alpha = 1$ for the first-order flux and $\alpha = \frac{2}{3}$ for the second-order flux. Hence, the force F_i also influences the stability criterion of the scheme.

A.2 Determination of the force

In order to determine the forces F_i, we need to solve (6.1) for the electric field. We approximate the electric field by $E_k^\ell \approx E(x_k, t_\ell)$. A second-order discretization for (6.1) is given by

$$\begin{aligned} \frac{E_{k+1}^* - E_k^*}{\Delta x} &= \frac{q_{k+1}^\ell + q_k^\ell}{2\varepsilon_0} \quad \text{for} \quad k = 0, ..., K-1, \\ \frac{E_0^* - E_K^*}{\Delta x} &= \frac{\rho_0^n + \rho_K^n}{2\varepsilon_0} \end{aligned} \tag{6.6}$$

which is calculated by a linear solver. In order to guarantee the zero-mean electrostatic condition (6.2), the electric field at time step $t = t_\ell$ is determined by

$$E_k^\ell = E_k^* - \frac{1}{K+1} \int E^* \, \mathrm{d}x. \tag{6.7}$$

Finally, the force reads

$$F_{i,k}^\ell = Z_i e E_k^\ell. \tag{6.8}$$

Having determined the force, we take a closer look to the time step restrictions of the entire scheme.

A.3 Time step restrictions

In order to guarantee stability and positivity, the advection in both, space and velocity, introduce restrictions for the size of the time step resulting in a CFL condition. The Courant number C needs to be chosen such that

$$0 < C \leq \alpha$$

where $\alpha = 1$ for the first-order numerical fluxes and $\alpha = \frac{2}{3}$ for the second-order numerical fluxes. In terms of positivity, the contributions with negative sign of the advection in veloctiy are to be added to the contributions with negative sign of the advection in space. Hence, considering only the convection parts, the time step size Δt for the update $t_\ell \to t_{\ell+1}$ is chosen to be

$$\Delta t = \Delta t_\ell = C \min_{q_1, i, k} \left\{ \frac{1}{\frac{|v_{q_1}^{(1)}|}{\Delta x} + \frac{|F_{i,k}^\ell|}{m_i \Delta v_i}} \right\} = C \min_{q_1, i, k} \left\{ \frac{\Delta x \, \Delta v_i}{\Delta v_i |v_{q_1}^{(1)}| + \Delta x \frac{|F_{i,k}^\ell|}{m_i}} \right\}. \tag{6.9}$$

Since the forces may change between the stages of the second-order schemes, but Δt is set during an entire time step, we make the assumption that the Courant number C comprises such changes.

Additionally, we we need to take care of the time step restrictions due to the positivity preservation during relaxation steps which is already described in Section 4.3.

Now, we are ready for numerical simulations.

A.4 Numerical tests

We perform initial numerical tests in order to illustrate the damping of a self-consistent electric field. In Section A.4.1, we start with the well-known Landau damping. In Section A.4.2, we study the behavior of the electric field when an additional relaxation process occurs.

A.4.1 Landau damping

We have introduced the phenomenon of Landau damping in Section 2.1.2. It describes the damping of the self-consistent electric field for the (linearized) Vlasov-Poisson system. More precisely, the electric energy

$$||E(x,t)||_{L_2} = \sqrt{\int E(x,t)^2 \,\mathrm{d}x} \tag{6.10}$$

decreases in time being converted to kinetic energy.

Ions are considered as a neutralizing background density $n_0 = 1$. We simulate electrons and the self-consistent electric field in a spatial domain $L = [0, 4\pi]$ with periodic boundary conditions. The distribution function of electrons follows the Vlasov equation representing the advection in phase space. Initially, we consider a slightly disturbed Maxwellian

$$f_e(x, \mathbf{v}, t = 0) = (1 + 0.01\,\cos(0.5\,x))\,\mathcal{M}[n_e, \mathbf{u}_e, T_e, m_e]$$

with

$$n_e = 1, \quad \mathbf{u}_e = 0, \quad T_e = 1, \quad m_e = 1,$$

mimicking a linearized system for our fully implemented equations. The charge number is $Z_e = -1$.

The simulation is run using a velocity grid with 24^3 nodes and a spatial mesh with 32 cells. We use the second-order IMEX RK scheme from Section 3.3.4 and the second-order spatial discretization in Section 3.4 with the limiter given in (3.71). The transport in $\mathbf{v}$ is discretized according to Sections A.1 and A.2. The time step is set to $\Delta t = 0.03$ which fulfils the CFL condition (6.9).

In Figure A.1, we provide the numerical result. The electric energy decays exponentially with the typical oscillations. The theory predicts a damping rate $\gamma = 0.153$ [FSB01] which is recovered by our simulation. The oscillations need to be studied in more detail. We suspect that the dissipative fluxes may lead to the shrinking of the amplitude of the oscillations. This needs to be verified in further tests.

Moreover, we observe a phenomenon which is known as Poincaré recurrence. This is a numerical phenomenon occuring for a periodic problem in space run on a uniform velocity grid [Son19]. To explain this incident, we consider the one-dimensional free streaming equation

$$\partial_t f + v\,\partial_x f = 0.$$

For a periodic problem in x with period L, we represent f by a Fourier series. The mode k of f obeys

$$\partial_t f_k - \frac{2\pi \mathrm{i} k}{L} v f_k = 0$$

with the solution $f_k(x,t) = e^{\frac{2\pi \mathrm{i} k}{L} vt} f_k(x,0)$. As the velocities are discretized by $v = q\Delta v$, we conclude that f_k is periodic in t with period $T = \frac{L}{k\Delta v}$. In our set-up, we have $L = 4\pi$ and $\Delta v = 0.5$. As the first mode dominates, the numerical result hits this recurrence time reasonably. By reducing Δv, the recurrence time gets larger and will not affect numerical results.

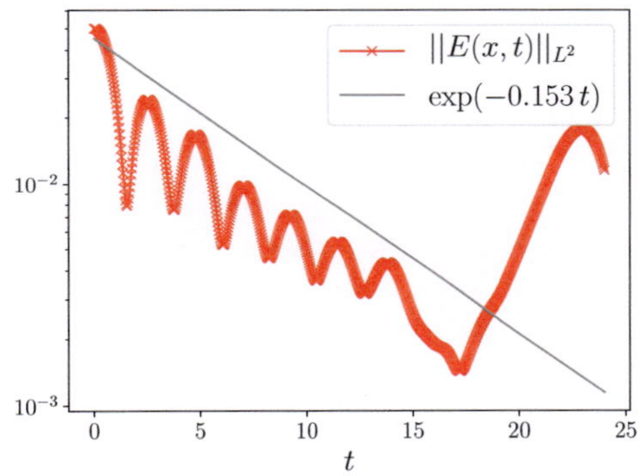

Figure A.1: Time evolution of the electric energy (6.10) for the Landau damping in Section A.4.1. The decay rate coincides with the theoretical prediction (gray line).

This test case only considers one species (electrons), whereas ions are treated via a background density. We are especially interested in multi-species equations where interactions are considered by BGK operators. Hence, the following tests handle the ions by evolving distribution functions.

A.4.2 Landau damping coupled with relaxation

We consider the full system (6.3) for two species where we use different collision frequencies. We vary the magnitude for

$$\hat{\nu}_{ij} = c \cdot \frac{n_j}{\delta_{ij} + \hat{v}} \tag{6.11}$$

by having different values of $c = 0, 10^{-1}, 10, 10^2, 10^3, 10^4$. Moreover, we implement the velocity-dependent collision frequency

$$\nu_{ij}(\mathbf{v}) = 10 \cdot \frac{n_j}{\delta_{ij} + |\mathbf{v} - \mathbf{u}_{\mathrm{mix}}|^3}. \tag{6.12}$$

The regularization parameter is $\delta_{ij} = 0.1 \cdot (\Delta v_{ij})^3$ where $\Delta v_{ij} = \frac{1}{4}\sqrt{T_{\text{mix}}/(2\mu_{ij}}$ and $\mu_{ij} = m_i m_j/(m_i + m_j)$. The averaged velocity $\hat{v}$ can be found in (2.102), respective (5.10).

We simulate ions (species 1) and electrons (species 2) in a spatial domain $L = [0, 4\pi]$ with periodic boundary conditions. The masses and charge numbers are

$$m_1 = m_2 = 1 \quad \text{and} \quad Z_1 = -Z_2 = 1.$$

The initial conditions read

$$\begin{aligned} f_1(x, \mathbf{v}, t = 0) &= \mathcal{M}[n_1, \mathbf{u}_1, T_1, m_1], \\ f_2(x, \mathbf{v}, t = 0) &= (1 + 0.01\, \cos(0.5\, x))\, \mathcal{M}[n_2, \mathbf{u}_2, T_2, m_2] \end{aligned}$$

with

$$\begin{aligned} n_1 &= 1, \quad \mathbf{u}_1 = (0.5, 0, 0)^\top, \quad T_1 = 1, \\ n_2 &= 1, \quad \mathbf{u}_2 = 0, \qquad\qquad\quad T_2 = 5. \end{aligned}$$

The simulations are run using a velocity grid with 48^3 nodes and 32 equally spaced cells in x. We use the second-order IMEX RK scheme from Section 3.3.4 with the limiter given in (3.71). The transport in $\mathbf{v}$ is discretized according to Sections A.1 and A.2. The time step $\Delta t = 0.0221$ is set according to the CFL condition (6.9).

In Figures A.2 and A.3, we compare the numerical results for the constant collision frequencies (6.11). The mean velocities and temperatures converge locally the faster the larger the collision frequencies are. In Figure A.2, we illustrate $||\mathbf{u}_1(x,t) - \mathbf{u}_2(x,t)||_\infty$ and $||T_1(x,t) - T_2(x,t)||_\infty$ as we consider a space inhomogeneous test case. These global quantities show another behavior when the collision frequencies become large ($c = 10^3, 10^4$). We observe a damping of the electric field for each case of c, see Figure A.3. The electric energy does not differ visibly for $c = 0$ and $c = 10^{-1}$. A small but considerable relaxation process ($c = 10, 10^2$) accelerates the damping of the electric field. However, a large value of $c = 10^3, 10^4$ corresponds to a slow decay of the electric field without oscillations. This phenomenon fits to the theoretical results in [Bau21] where it is important to assume small collision frequencies in order to prove the damping of the electric field. It would be worth to study the effect of the size of the collision frequency in more detail.

In Figures A.4 and A.5, we compare the result for the velocity-dependent collision frequency (6.12) to the corresponding constant one, (6.11) with $c = 10$. As we already realized in previous test cases, the relaxation process of the mean velocities and temperatures is significantly different when velocity-dependent collision frequencies (of comparable size) are encountered, see Figure A.4. In Figure A.5, we observe that also the damping of the electric energy is affected when using velocity-dependent collision frequencies.

To summarize, the coupling to a self-consistent electric field leads to interesting phenomena and asks for more numerical tests. The provided simulations serve as a first sample.

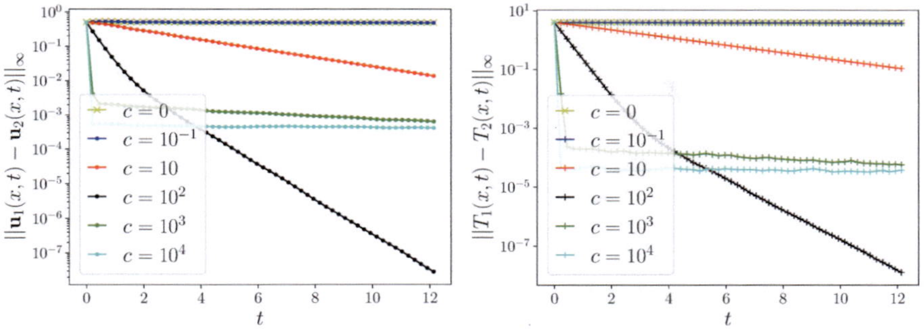

(a) mean velocities **(b)** temperatures

Figure A.2: Illustration of the time behavior of the mean velocities and temperatures for the test case in Section A.4.2. The constant collision frequency $\hat{\nu}_{ij}$ in (6.11) takes different values for the factor c. The larger c the faster is the local convergence in every cell. This does not translate one-to-one to the global behavior of $||\mathbf{u}_1(x,t) - \mathbf{u}_2(x,t)||_\infty$ and $||T_1(x,t) - T_2(x,t)||_\infty$.

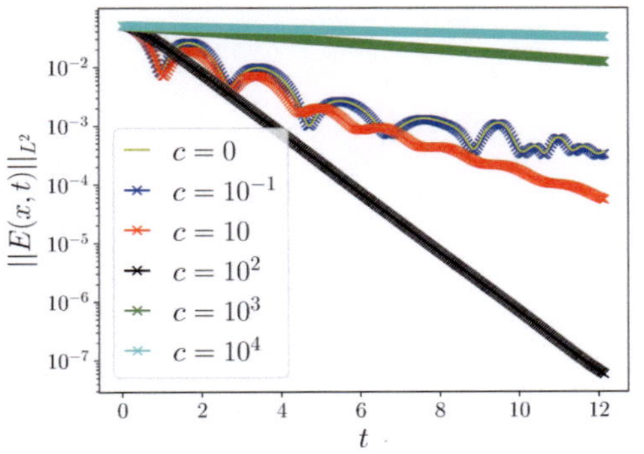

Figure A.3: Damping of the electric field for the test case in Section A.4.2. The constant collision frequency $\hat{\nu}_{ij}$ in (6.11) takes different values for the factor c. The curves for $c = 0$ and $c = 10^{-1}$ coincide. For small collision frequencies, the phenomenon of Landau damping is still visible. When the collision frequencies become larger, the relaxation process predominates resulting in a significantly different decay of the electric energy.

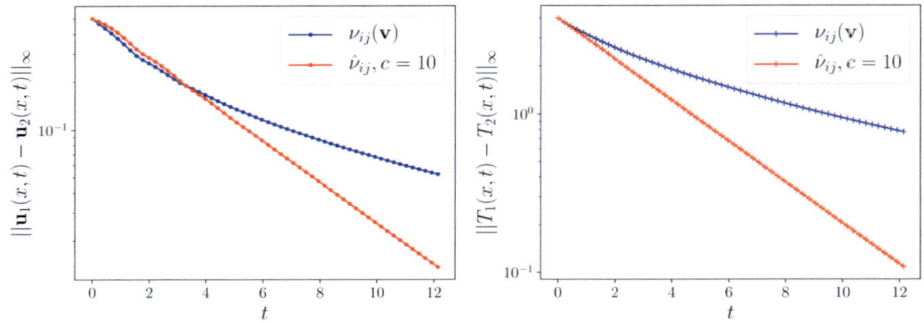

(a) mean velocities

(b) temperatures

Figure A.4: Illustration of the time behavior of the mean velocities and temperatures for the test case in Section A.4.2 where we compare between the result for the constant collision frequency $\hat{\nu}_{ij}$ in (6.11) using the factor $c = 10$ and the result for the velocity-dependent collision frequency $\nu_{ij}(\mathbf{v})$ in (6.12). The relaxation process is obviously different.

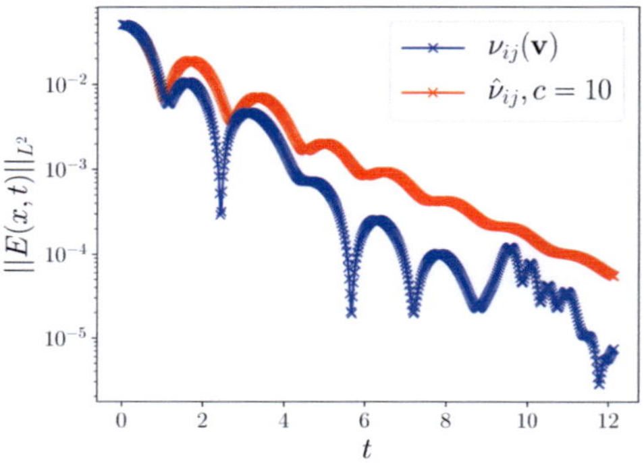

Figure A.5: Damping of the electric field for the test case in Section A.4.2 where we compare between the result for the constant collision frequency $\hat{\nu}_{ij}$ in (6.11) using the factor $c = 10$ and the result for the velocity-dependent collision frequency $\nu_{ij}(\mathbf{v})$ in (6.12). The decay of the electric energy shows a significantly different behavior.

B General implicit solver for 3 species

The presented scheme is based on the general implicit solver in Section 3.3.5. We shortly illustrate that this method can be extended to more than 2 species in a straight-forward way. Let us consider 3 species described by distribution functions f_1, f_2 and f_3. The transport operators act on the individual species such that only the relaxation process is of interest here.

As in Section 3.3.5, we write the implicit updates of the distribution functions in a generic steady state form

$$f_i = c_i G_i + c_i \gamma \Delta t(\nu_{ii}\mathcal{A}_{ii,\tau_i} + \nu_{ij}\mathcal{A}_{ij,\tau_i} + \nu_{ik}\mathcal{A}_{ik,\tau_i}) \tag{6.13}$$

for $i, j, k \in \{1, 2, 3\}$, each of i, j, k distinct, where $\mathcal{A}_{ii,\tau_i}$, $\mathcal{A}_{ij,\tau_i}$ and $\mathcal{A}_{ik,\tau_i}$ are the unique target functions associated to f_i,

$$c_i = \frac{1}{1 + \gamma\Delta t(\nu_{ii} + \nu_{ij} + \nu_{ik})}, \tag{6.14}$$

and G_i is a known function. We aim to express $\mathcal{A}_{ii,\tau_i}$, $\mathcal{A}_{ij,\tau_i}$ and $\mathcal{A}_{ik,\tau_i}$ as functions of G_i, G_j and G_k so that (6.13) provides an explicit update formula for f_i.

We apply the conservation properties to (6.13). An analogous calculation as in the 2-species case leads to

$$\begin{aligned}
&\int c_1 \left(\nu_{11}\mathcal{A}_{11,\tau_1} + \nu_{12}\mathcal{A}_{12,\tau_1} + \nu_{13}\mathcal{A}_{13,\tau_1}\right) \mathbf{a}_1 \,\mathrm{d}\mathbf{w} \\
&+ \int c_2 \left(\nu_{21}\mathcal{A}_{21,\tau_2} + \nu_{22}\mathcal{A}_{22,\tau_2} + \nu_{23}\mathcal{A}_{23,\tau_2}\right) \mathbf{a}_2 \,\mathrm{d}\mathbf{w} \\
&+ \int c_3 \left(\nu_{31}\mathcal{A}_{31,\tau_3} + \nu_{32}\mathcal{A}_{32,\tau_3} + \nu_{33}\mathcal{A}_{33,\tau_3}\right) \mathbf{a}_3 \,\mathrm{d}\mathbf{w} \\
&\quad = \int c_1 \left(\nu_{11} + \nu_{12} + \nu_{13}\right) G_1 \mathbf{a}_1(\mathbf{w}) \,\mathrm{d}\mathbf{w} + \int c_2 \left(\nu_{21} + \nu_{22} + \nu_{23}\right) G_2 \mathbf{a}_2(\mathbf{w}) \,\mathrm{d}\mathbf{w} \\
&\quad + \int c_3 \left(\nu_{31} + \nu_{32} + \nu_{33}\right) G_3 \mathbf{a}_3(\mathbf{w}) \,\mathrm{d}\mathbf{w},
\end{aligned} \tag{6.15}$$

which provides a set of constraints to determine the target functions from the given data.

The constraints in (6.15) represent first-order optimality conditions associated to the minimization of the convex function

$$\begin{aligned}
&\varphi_{\mathrm{tot}}(\boldsymbol{\alpha}_1, \boldsymbol{\alpha}_2, \boldsymbol{\alpha}_3, \boldsymbol{\alpha}_{12}, \boldsymbol{\alpha}_{13}, \boldsymbol{\alpha}_{23}) \\
&\qquad = \varphi_1(\boldsymbol{\alpha}_1) + \varphi_2(\boldsymbol{\alpha}_2) + \varphi_3(\boldsymbol{\alpha}_3) + \varphi(\boldsymbol{\alpha}_{12}) + \varphi(\boldsymbol{\alpha}_{13}) + \varphi(\boldsymbol{\alpha}_{23})
\end{aligned}$$

with

$$\varphi_i(\boldsymbol{\alpha}_i) = \int c_i \nu_{ii} \mathfrak{h}_{\tau_i}[\mathcal{A}_{ii,\tau_i}] \,\mathrm{d}\mathbf{w} + \boldsymbol{\mu}_{ii} \cdot \boldsymbol{\alpha}_i$$

and

$$\varphi(\boldsymbol{\alpha}_{ij}) = \int \left(c_i\nu_{ij}\mathfrak{h}_{\tau_i}[\mathcal{A}_{ij,\tau_i}] + c_j\nu_{ji}\mathfrak{h}_{\tau_j}[\mathcal{A}_{ji,\tau_j}]\right) \mathrm{d}\mathbf{w} + \boldsymbol{\mu}_{ij} \cdot \boldsymbol{\alpha}_{ij}.$$

This formulation already suggests that the minimization can be decoupled. As in the 2-species case, we have

$$\mathfrak{h}_{\tau_i}[\mathcal{A}_{ij,\tau_i}] = \frac{\log(1-\tau_i\mathcal{A}_{ij,\tau_i})}{\tau_i} = \begin{cases} -\mathcal{A}_{ij,\tau_i} & \text{for } \tau_i = 0, \\ \log(1-\mathcal{A}_{ij,1}) & \text{for } \tau_i = +1, \\ -\log(1+\mathcal{A}_{ij,-1}) & \text{for } \tau_i = -1. \end{cases}$$

Moreover, $\boldsymbol{\alpha}_i = (\alpha_i^{(0)}, \boldsymbol{\alpha}_i^{(1)}, \alpha_i^{(2)})^\top$;

$$\boldsymbol{\mu}_{ii} = \begin{pmatrix} \mu_{ii}^{(0)} \\ \boldsymbol{\mu}_{ii}^{(1)} \\ \mu_{ii}^{(2)} \end{pmatrix} = \int c_i\nu_{ii}G_i\mathbf{a}_i(\mathbf{w})\,\mathrm{d}\mathbf{w}$$

for $i = 1, 2, 3$; for $i \neq j : \boldsymbol{\alpha}_{ij} = (\alpha_{ij}^{(0)}, \alpha_{ji}^{(0)}, \boldsymbol{\alpha}_{ij}^{(1)}, \alpha_{ij}^{(2)})^\top$; and

$$\boldsymbol{\mu}_{ij} = \begin{pmatrix} \mu_{ij}^{(0)} \\ \mu_{ji}^{(0)} \\ \boldsymbol{\mu}_{ij}^{(1)} \\ \mu_{ij}^{(2)} \end{pmatrix} = \int \left[\begin{pmatrix} \mathrm{a}_i^{(0)}(\mathbf{w}) \\ 0 \\ \mathbf{a}_i^{(1)}(\mathbf{w}) \\ \mathrm{a}_i^{(2)}(\mathbf{w}) \end{pmatrix} c_i\nu_{ij}G_i + \begin{pmatrix} 0 \\ \mathrm{a}_j^{(0)}(\mathbf{w}) \\ \mathbf{a}_j^{(1)}(\mathbf{w}) \\ \mathrm{a}_j^{(2)}(\mathbf{w}) \end{pmatrix} c_j\nu_{ji}G_j \right] \mathrm{d}\mathbf{w}.$$

C Pseudo codes

The main code has a simple structure, see Algorithm 1. The update of the distribution functions depends on the chosen time and space discretization. In Algorithms 2–5, we give the pseudo codes for the first-order splitting scheme, the first-order IMEX RK scheme, the Strang splitting and the second-order IMEX RK scheme. In Algorithm 6, we provide the pseudo code for the transport step needed for the splitting schemes. The code's core is represented by the relaxation process, see Algorithm 7. We conclude with Algorithm 8 where the update of the force according to a self-consistent electric field is described.

Algorithm 1 The structure of the main code.

```
set initial and boundary conditions
create grids (see Sections 3.4 and 3.5)
initialize distribution functions
determine initial target functions (see Remark 3.3.2)
if force active then
   determine initial force (see Algorithm 8)
end if
set time step according to CFL condition (see Sections 4.3 and A.3)
for iterations do
   if force active or positivity violated then
      calculate new time step size (see Sections 4.3 and A.3)
   end if
   update distribution functions (Algorithms 2–5)
end for
```

Algorithm 2 Update of distribution functions according to first-order splitting (Section 3.3.1): The scheme consists of a first-order relaxation step, followed by a first-order transport step.

```
first-order relaxation step (Algorithm 7)
first-order transport step (Algorithm 6)
```

Algorithm 3 Update of distribution functions according to first-order IMEX RK (Section 3.3.2): The numerical fluxes are needed in order to compute the input data for the first-order relaxation step.

```
if force active then
    update force (see Algorithm 8)
end if
calculate first-order numerical fluxes (see Sections 3.4.1 and A.1)
first-order relaxation step (Algorithm 7)
```

Algorithm 4 Update of distribution functions according to Strang splitting (Section 3.3.3): The update starts and ends with a second-order transport step. In between, the target functions need to be updated because they are needed for the input data of the relaxation step.

```
second-order transport step with Δt/2 (Algorithm 6)
update target functions (see Remark 3.3.2)
second-order relaxation step with Δt (Algorithm 7)
second-order transport step with Δt/2 (Algorithm 6)
```

Algorithm 5 Update of distribution functions according to second-order IMEX RK (Section 3.3.4): Both stages begin with the numerical fluxes in order to compute the input data for the second-order relaxation step.

```
for stages 1 and 2 do
    if force active then
        update force (see Algorithm 8)
    end if
    calculate second-order numerical fluxes (see Sections 3.4.1 and A.1)
    second-order relaxation step (Algorithm 7)
end for
```

Algorithm 6 The transport step (Section 3.4) follows either the Forward Euler method (3.2) or Heun's method (3.4).

```
for stages do
    if force active then
        update force (see Algorithm 8)
    end if
    calculate numerical fluxes (see Sections 3.4.1 and A.1)
    update distribution functions e.g. by a combination of (3.73) and (6.4)
end for
```

Algorithm 7 The structure of the relaxation step (Section 3.3.5) is illustrated. The input data possibly takes numerical fluxes and target functions evaluated at previous steps/stages into account. This depends on the chosen time and space discretization. Afterwards, varying collision frequencies are updated. Due to Assumption 2.2.11, the collision frequencies remain untouched by the relaxation process which guarantees the conservation properties at the discrete level. The potential functions are minimized by a Newton algorithm which requires the evaluation of gradients and Hessians. This procedure yields the values of the target functions at the next time step, followed by the update of the distribution functions. Eventually, the positivity of the distribution functions is checked (for the second-order schemes).

```
calculate input data (3.56) and (3.57) (dependent on time discretization)
update collision frequencies (see Sections 2.2.3 and 3.3)
minimize potential functions (3.58) and (3.59) by Newton's method which deter-
    mines new values of target functions
update distribution functions by (3.50)
check positivity (see Section 4.2)
```

Algorithm 8 The force is determined according to Section A.2. The crucial step is to solve a system of linear equations which determines the self-consistent electric field.

```
calculate charge density
calculate electric field by solving (6.6) via a linear solver
enforce zero-mean electrostatic condition by (6.7)
calculate force by (6.8)
```

D Units

In the presented models, we do not emphasize the use of units which is often done in mathematics. However, for physical reasons units should not be disregarded.

One option to avoid unit mess is to use dimensionless equations. Once these are derived, no more attention for units is needed. Additionally, one only has to deal with mass ratios, not with individual, possibly very small masses.

In this thesis, we have not considered dimensionless equations, but we made use of a normalization, see Remark 5.2.1. We further mention a few things regarding units for our equations.

D.1 CGS system

There exist several unit systems. In plasma and astrophysics, people mainly use the CGS system. The basic units are

- $[s] = \mathrm{cm}$ for path length,
- $[m] = \mathrm{g}$ for mass,
- $[t] = \mathrm{s}$ for time,

which explains the system's name. Additionally, the system is augmented by electrostatic units. The base energy unit is

- $[E] = \mathrm{erg} = \frac{\mathrm{g} \cdot \mathrm{cm}^2}{\mathrm{s}^2}$.

Nevertheless, termperatures are usually expressed in another energy unit:

- $[T] = \mathrm{eV}$.

This means that a conversion between the temperature and the base energy unit is required (similar to the SI unit system) which is often denoted by a Boltzmann constant k_B, i.e.

$$3n \underbrace{k_B}_{\mathrm{erg/eV}} \underbrace{T}_{\mathrm{eV}} = \int \underbrace{m|\mathbf{v} - \mathbf{u}|^2}_{\mathrm{erg}} f(\mathbf{v}) \,\mathrm{d}\mathbf{v}. \tag{6.16}$$

Accordingly, the typical variables in kinetic simulations have the following units:

- number density: $[n] = 1/\,\mathrm{cm}^3$,
- velocity: $[\mathbf{v}] = \mathrm{cm}/\,\mathrm{s}$,
- temperature: $[T] = \mathrm{eV} = 1.60218 \times 10^{-12}\, \frac{\mathrm{g\,cm}^2}{\mathrm{s}^2}$.

More information including constants in CGS units can be found in Appendix E. For instance, the vacuum permitivity reads $\varepsilon_0 = 1/4\pi$. This simple representation illustrates why the CGS system is useful. When the CGS system is used in the literature, $4\pi\varepsilon_0$ is often set to 1 already in the formulas.

D.2 Units in the quantum case

The formulae in Section 2.3 suggest $[f] = 1$. Though, the distribution function is a density in phase space; accordingly, we require

$$[f] = \frac{1}{[\mathbf{x}][\mathbf{p}]}.$$

Replacing

$$f \mapsto \frac{f}{h^3} \cdot \frac{m^3}{\beta}$$

in all formulae corrects this issue. Here, m is the mass, h is the Planck constant and β is the statistical weight of a particle (being the number of independent quantum states with the same internal energy). We additionally need to replace

$$\tau \mapsto h^3 \frac{\beta}{m^3} \tau.$$

For more information, see [CC70, Chapter 17].

E List of symbols

We provide a collection of used notation and give the CGS unit when possible. Firstly, we present an overview over used indeces and notation for coordinates.

symbol	quantity/description	CGS unit	index, sub/superscript
	number of species		$i, j \downarrow$
	type of species (classic, fermion, boson)		$\tau \downarrow$
	component r of vector		$(r) \uparrow$
	coordinates		
t	time coordinate	s	$\ell \uparrow$ $*$ (splitting step) $\uparrow$ $[\cdot]$ (stage) $\uparrow$
$\mathbf{x}$	space coordinate	cm	$k \downarrow$
$(\mathbf{w} =)\mathbf{v}$	velocity coordinate	$\frac{\mathrm{cm}}{\mathrm{s}}$	$\mathbf{q} \downarrow$
$(\mathbf{w} =)\mathbf{p}$	momentum coordinate	$\mathrm{g}\frac{\mathrm{cm}}{\mathrm{s}}$	$\mathbf{q} \downarrow$
$\mathbf{a}$	vector summarizing $1, \mathbf{w}, \lvert\mathbf{w}\rvert^2$		

The following table presents the symbols used for operators, distribution functions and correlated quantities, macroscopic and physical quantities.

symbol	quantity/description	CGS unit
	operators and more	
$\mathcal{Q}$	interaction operator	
$\mathcal{R}$	relaxation operator	
$\mathcal{T}$	transport operator	
$\mathscr{F}$	numerical fluxes	
	distribution functions	
f	distribution function (classic)	$\frac{1}{[\mathbf{x}]^3[\mathbf{v}]^3} = \frac{1}{\mathrm{cm}^3}\frac{\mathrm{s}^3}{\mathrm{cm}^3}$
f	distribution function (quantum)	$\frac{1}{[\mathbf{x}]^3[\mathbf{p}]^3} = \frac{1}{\mathrm{cm}^3}\frac{\mathrm{s}^3}{\mathrm{g}^3\,\mathrm{cm}^3}$

$\mathcal{A}$	target function/attractor	
$\mathcal{M}$	Maxwell distribution function	$\frac{1}{[\mathbf{x}]^3[\mathbf{v}]^3} = \frac{1}{\mathrm{cm}^3}\frac{\mathrm{s}^3}{\mathrm{cm}^3}$
$\mathcal{G}$	classical target function for $\nu(\mathbf{v})$	$\frac{1}{[\mathbf{x}]^3[\mathbf{v}]^3} = \frac{1}{\mathrm{cm}^3}\frac{\mathrm{s}^3}{\mathrm{cm}^3}$
$\mathcal{E}$	quantum target function	$\frac{1}{[\mathbf{x}]^3[\mathbf{p}]^3} = \frac{1}{\mathrm{cm}^3}\frac{\mathrm{s}^3}{\mathrm{g}^3\,\mathrm{cm}^3}$
$\mathcal{F}$	Fermi-Dirac distribution function	$\frac{1}{[\mathbf{x}]^3[\mathbf{p}]^3} = \frac{1}{\mathrm{cm}^3}\frac{\mathrm{s}^3}{\mathrm{g}^3\,\mathrm{cm}^3}$
$\mathcal{B}$	Bose-Einstein distribution function	$\frac{1}{[\mathbf{x}]^3[\mathbf{p}]^3} = \frac{1}{\mathrm{cm}^3}\frac{\mathrm{s}^3}{\mathrm{g}^3\,\mathrm{cm}^3}$
$\boldsymbol{\lambda}$	Lagrange multiplier; parameter in $\mathcal{A}$	
$\boldsymbol{\mu}$	given input data	$([\rho], [n\mathbf{u}], [E])^\top$
	macroscopic quantities	
ρ	mass density	$\frac{[m]}{[\mathbf{x}]^3} = \frac{\mathrm{g}}{\mathrm{cm}^3}$
n	number density	$\frac{1}{[\mathbf{x}]^3} = \frac{1}{\mathrm{cm}^3}$
$\mathbf{u}$	mean velocity	$[\mathbf{v}] = \frac{\mathrm{cm}}{\mathrm{s}}$
T	(kinetic) temperature	eV
ϑ	(physical) temperature	eV
$E(e)$	(internal) energy density	$\frac{[m][\mathbf{v}]^2}{[\mathbf{x}]^3} = \mathrm{g}\frac{\mathrm{cm}^2}{\mathrm{s}^2}\frac{1}{\mathrm{cm}^3}$
$\mathbf{Q}$	energy flux	
$\mathbb{P}$	pressure tensor	$\frac{\mathrm{g}}{\mathrm{cm\,s}}$
p	pressure	$\frac{\mathrm{g}}{\mathrm{cm\,s}}$
q	electric charge density	$\frac{\sqrt{\mathrm{eV\,cm}}}{\mathrm{cm}^3}$
$\mathbf{j}$	electric current	$\frac{\sqrt{\mathrm{eV\,cm}}\,\mathrm{cm}}{\mathrm{cm}^3\,\mathrm{s}}$
$\mathcal{H}$	entropy density	
h	integrand for entropy density	
D	entropy dissipation	
φ	potential function	
$\mathfrak{H}$	'discrete' entropy density in scheme	
$\mathfrak{h}$	integrand for 'discrete' entropy density	

physical quantities		
m	mass	g
μ	reduced mass	g
ν	collision frequency	$\frac{1}{\mathrm{s}}$
$\mathbf{E}$	electric field	$10^{-2}\,\frac{\text{Volt}}{\mathrm{cm}}$
$\mathbf{B}$	magnetic field	$10^{3}\,\frac{\mathrm{g}}{\mathrm{s}^2\,\text{Ampère}}$
$\mathbf{F}$	force	$\mathrm{g}\,\frac{\mathrm{cm}}{\mathrm{s}^2}$
$\log\Lambda$	Coulomb logarithm	1
λ_D	Debye length/screening length	cm
λ_{dB}	deBroglie wave length	cm
b_{90}	distance of closest approach	cm
Z	charge number	1
Kn	Knudsen number	1
Pr	Prandtl number	1

We conclude with a summary of some physical constants.

symbol	quantity/description	value of constant	CGS unit
physical constants			
e^2	square of elementary charge	$1.44 \cdot 10^{-7}$	eV cm
k_B	Boltzmann constant	$1.60218 \cdot 10^{-12}$	$\frac{\mathrm{erg}}{\mathrm{eV}}$
h	Planck constant	$4.135667696 \cdot 10^{-15}$	$[\mathbf{p}][\mathbf{x}] = \mathrm{eV\,s}$
ε_0	vacuum permittivity	$\frac{1}{4\pi}$	1
c	speed of light in vacuum	$2.99792458 \cdot 10^{10}$	$\frac{\mathrm{cm}}{\mathrm{s}}$
μ_0	vacuum permeability	$\frac{1}{\varepsilon_0 c^2}$	$\frac{\mathrm{s}^2}{\mathrm{cm}^2}$

F Glossary of abbreviations

cm-g-s **(CGS)** unit system. 3, 57, 136, 141, 147, 150, 178, 180

AP asymptotic-preserving. 19, 88, 89, 99, 162

BBGKY Bogoliubov–Born–Green–Kirkwood–Yvon. 11–13

BGK Bhatnagar–Gross–Krook. ix, 2, 5, 17, 24, 32–36, 38, 41–43, 45, 46, 54–56, 66, 67, 71, 73, 79, 81, 86, 89, 90, 94, 99–102, 114, 115, 119, 131, 139, 147, 154, 161, 162, 166, 169

CFL Courant-Friedrichs-Lewy. 86, 100, 102, 113, 117–120, 122, 123, 128, 132, 147, 148, 150, 151, 153, 166, 168, 170

DIRK diagonally implicit Runge-Kutta. 82, 88

ES-BGK ellipsoidal statistical BGK. 35

FD finite difference. 90

FEM finite element. 90, 92

FV finite volume. 90, 92, 99, 102, 113, 124, 147, 161, 166

GSA globally stiffly accurate. 88, 108

ICF inertial confinement fusion. 2, 150

IMEX implicit-explicit. ix, 86–90, 101, 105, 106, 108, 117, 119, 120, 132, 133, 135–137, 143, 147, 148, 150, 151, 153, 157, 168, 170, 175

ITER International Thermonuclear Experimental Reactor. 2

JET Joint European Torus. 2

KPP Klingenberg-Pirner-Puppo. 36, 42, 43, 79, 100, 134

LHS left-hand side. 12, 13, 20, 21, 41, 74, 165

ODE ordinary differential equation. 81, 83, 86, 89, 91

PDE partial differential equation. 1, 22, 24, 83, 87, 89, 92, 102, 106, 165

RHS right-hand side. 12, 13, 20, 28, 34–36, 41, 117, 124, 147

RK Runge-Kutta. 81–85, 87–90, 101, 106, 108, 117, 119, 120, 132, 133, 135–137, 143, 147, 148, 150, 151, 153, 157, 168, 170, 175

SL semi-Lagrangian. 93, 163

SSP strong-stability-preserving. 88–90, 119, 120

Bibliography

[AAC16] Franz Achleitner, Anton Arnold, and Eric A. Carlen. On Linear Hypocoercive BGK Models. In Patrícia Gonçalves and Ana Jacinta Soares, editors, From Particle Systems to Partial Differential Equations III, pages 1–37, Cham, 2016. Springer International Publishing.

[AAP02] Pierre Andries, Kazuo Aoki, and Benoit Perthame. A Consistent BGK-Type Model for Gas Mixtures. Journal of Statistical Physics, 106(5):993–1018, 2002. doi:10.1023/A:1014033703134.

[Abr07] Rafail V. Abramov. An improved algorithm for the multidimensional moment-constrained maximum entropy problem. Journal of Computational Physics, 226:621–644, 2007.

[AHOT14] Graham W. Alldredge, Cory D. Hauck, Dianne P. O'Leary, and André L. Tits. Adaptive change of basis in entropy-based moment closures for linear kinetic equations. Journal of Computational Physics, 258:489 – 508, 2014/02/01 2014. doi:10.1016/j.jcp.2013.10.049.

[AHT12] Graham W. Alldredge, Cory D. Hauck, and André L. Tits. High-Order Entropy-Based Closures for Linear Transport in Slab Geometry II: A Computational Study of the Optimization Problem. SIAM Journal on Scientific Computing, 34(4):B361–B391, 2012. doi:10.1137/11084772X.

[ALPP00] Pierre Andries, Patrick Le Tallec, Jean-Philippe Perlat, and Benoît Perthame. The Gaussian-BGK model of Boltzmann equation with small Prandtl number. European Journal of Mechanics - B/Fluids, 19(6):813–830, 2000. doi:10.1016/S0997-7546(00)01103-1.

[And89] J. D. Anderson. Hypersonic and high temperature gasdynamics. McGraw-Hill, 1989.

[AP12] Alessandro Alaia and Gabriella Puppo. A hybrid method for hydrodynamic-kinetic flow – Part II – Coupling of hydrodynamic and kinetic models. Journal of Computational Physics, 231(16):5217–5242, 2012. doi:10.1016/j.jcp.2012.02.022.

[ARS97] Uri M. Ascher, Steven J. Ruuth, and Raymond J. Spiteri. Implicit-explicit Runge-Kutta methods for time-dependent partial differential equations. Applied Numerical Mathematics, 25(2):151–167, 1997. Special Issue on Time Integration. doi:10.1016/S0168-9274(97)00056-1.

[Asi08] Pietro Asinari. Asymptotic Analysis of Multiple-Relaxation-Time Lattice Boltzmann Schemes for Mixture Modeling. Computers & Mathematics with Applications, 55(7):1392–1407, April 2008. doi:10.1016/j.camwa.2007.08.006.

[Atk89] Kendall E Atkinson. An introduction to numerical analysis. John Wiley & Sons, second edition, 1989.

[BAA+21a] Satish Balay, Shrirang Abhyankar, Mark F. Adams, Steven Benson, Jed Brown, Peter Brune, Kris Buschelman, Emil Constantinescu, Lisandro Dalcin, Alp Dener, Victor Eijkhout, William D. Gropp, Václav Hapla, Tobin Isaac, Pierre Jolivet, Dmitry Karpeev, Dinesh Kaushik, Matthew G. Knepley, Fande Kong, Scott Kruger, Dave A. May, Lois Curfman McInnes, Richard Tran Mills, Lawrence Mitchell, Todd Munson, Jose E. Roman, Karl Rupp, Patrick Sanan, Jason Sarich, Barry F. Smith, Stefano Zampini, Hong Zhang, Hong Zhang, and Junchao Zhang. PETSc/TAO Users Manual. Technical Report ANL-21/39 - Revision 3.16, Argonne National Laboratory, 2021.

[BAA+21b] Satish Balay, Shrirang Abhyankar, Mark F. Adams, Steven Benson, Jed Brown, Peter Brune, Kris Buschelman, Emil M. Constantinescu, Lisandro Dalcin, Alp Dener, Victor Eijkhout, William D. Gropp, Václav Hapla, Tobin Isaac, Pierre Jolivet, Dmitry Karpeev, Dinesh Kaushik, Matthew G. Knepley, Fande Kong, Scott Kruger, Dave A. May, Lois Curfman McInnes, Richard Tran Mills, Lawrence Mitchell, Todd Munson, Jose E. Roman, Karl Rupp, Patrick Sanan, Jason Sarich, Barry F. Smith, Stefano Zampini, Hong Zhang, Hong Zhang, and Junchao Zhang. PETSc Web page, 2021. URL: https://petsc.org/.

[Bab98] Hans Babovsky. Die Boltzmann-Gleichung. B.G. Teubner, Stuttgart Leipzig, 1998.

[Bal60] R. Balescu. Irreversible Processes in Ionized Gases. The Physics of Fluids, 3(1):52–63, 1960. doi:10.1063/1.1706002.

[Bau21] Lena Baumann. On Landau damping coupled with relaxation for the Vlasov-Poisson-BGK system. Master's thesis, Julius-Maximilians-Universität Würzburg, 2021. URL: https://ifm.mathematik.uni-wuerzburg.de/~klingen/ewExternalFiles/Baumann%20Lena_Masters%20thesis.pdf.

[BBG+18] Alexander V. Bobylev, Marzia Bisi, Maria Groppi, Giampiero Spiga, and Irina F. Potapenko. A general consistent BGK model for gas mixtures. Kinetic & Related Models, 11(6):1377–1393, 2018.

[BCGR21] Sebastiano Boscarino, Seung Yeon Cho, Maria Groppi, and Giovanni Russo. BGK models for inert mixtures: Comparison and applications. Kinetic & Related Models, 14(5):895–928, 2021.

[BCK21] Jonas P. Berberich, Praveen Chandrashekar, and Christian Klingenberg. High order well-balanced finite volume methods for multi-dimensional systems of hyperbolic balance laws. Computers & Fluids, 219:104858, 2021. doi:10.1016/j.compfluid.2021.104858.

[BCNS12] Abdelghani Bellouquid, Juan Calvo, Juanjo Nieto, and Juan Soler. On the Relativistic BGK-Boltzmann Model: Asymptotics and Hydrodynamics. Journal of Statistical Physics, 149, 05 2012. doi:10.1007/s10955-012-0600-0.

[BCR22] Sebastiano Boscarino, Seung Yeon Cho, and Giovanni Russo. A local velocity grid conservative semi-Lagrangian schemes for BGK model. Journal of Computational Physics, 460:111178, 2022. doi:10.1016/j.jcp.2022.111178.

[BCS11] Blanca Ayuso, José A. Carrillo, and Chi-Wang Shu. Discontinuous Galerkin methods for the one-dimensional Vlasov-Poisson system. Kinetic & Related Models, 4(4):955–989, 2011.

[Ber21] Jonas Philipp Berberich. Fluids in Gravitational Fields — Well-Balanced Modifications for Astrophysical Finite-Volume Codes. doctoralthesis, Universität Würzburg, 2021. doi:10.25972/OPUS-21967.

[BG46] Max Born and H. S. Green. A general kinetic theory of liquids I. The molecular distribution functions. Proceedings of the Royal Society of London. Series A. Mathematical and Physical Sciences, 188(1012):10–18, 1946. doi:10.1098/rspa.1946.0093.

[BGK54] P. L. Bhatnagar, E. P. Gross, and M. Krook. A Model for Collision Processes in Gases. I. Small Amplitude Processes in Charged and Neutral One-Component Systems. Physical Review, 94:511–525, May 1954. doi:10.1103/PhysRev.94.511.

[BGMS97] Satish Balay, William D. Gropp, Lois Curfman McInnes, and Barry F. Smith. Efficient management of parallelism in object oriented numerical software libraries. In E. Arge, A. M. Bruaset, and H. P. Langtangen, editors, Modern Software Tools in Scientific Computing, pages 163–202. Birkhäuser Press, 1997.

[BHLPD+15] L. F. Berzak Hopkins, S. Le Pape, L. Divol, N. B. Meezan, A. J. Mackinnon, D. D. Ho, O. S. Jones, S. Khan, J. L. Milovich, J. S. Ross, P. Amendt, D. Casey, P. M. Celliers, A. Pak, J. L. Peterson, J. Ralph, and J. R. Rygg. Near-vacuum hohlraums for driving fusion implosions with high density carbon ablators. Physics of Plasmas, 22(5):056318, 2015.

[BIP14] Florian Bernard, Angelo Iollo, and Gabriella Puppo. A Local Velocity Grid Approach for BGK Equation. Communications in Computational Physics, 16(4):956–982, 2014. doi:10.4208/cicp.291013.240314a.

[BIP15] Florian Bernard, Angelo Iollo, and Gabriella Puppo. Accurate Asymptotic Preserving Boundary Conditions for Kinetic Equations on Cartesian Grids. Journal of Scientific Computing, 65:735–766, 2015.

[Bir80] G. Bird. Monte-carlo simulation in an engineering context. In 12th International Symposium on rarefied gas dynamics, 1980.

[Bir94] G. Bird. Molecular Gas Dynamics and the Direct Simulation of Gas Flows. 1994.

[BKPY21] Gi-Chan Bae, Christian Klingenberg, Marlies Pirner, and Seok-Bae Yun. BGK model of the multi-species Uehling-Uhlenbeck equation. Kinetic & Related Models, 14(1):25–44, 2021.

[BKYP22] G. Bae, C. Klingenberg, S. Yun, and M. Pirner. BGK model for a multi-component gas near a global Maxwellian. submitted, 2022. arXiv:2201.01611.

[BLM08] Mounir Bennoune, Mohammed Lemou, and Luc Mieussens. Uniformly stable numerical schemes for the Boltzmann equation preserving the compressible Navier–Stokes asymptotics. Journal of Computational Physics, 227(8):3781–3803, 2008. doi:10.1016/j.jcp.2007.11.032.

[BM14] S. Brull and L. Mieussens. Local discrete velocity grids for deterministic rarefied flow simulations. Journal of Computational Physics, 266:22–46, 2014. doi:10.1016/j.jcp.2014.01.050.

[Bog46] N. N. Bogoliubov. Kinetic Equations. Journal of Experimental and Theoretical Physics, 16(8):691–702, 1946.

[Bol09] Ludwig Boltzmann. Weitere Studien über das Wärmegleichgewicht unter Gas-molekülen. Wissenschaftliche Abhandlungen Band I, pages 316—-402, 1909.

[BP93] François Bouchut and Benoit Perthame. A BGK model for small Prandtl number in the Navier-Stokes approximation. Journal of Statistical Physics, 71:191–207, 01 1993. doi:10.1007/BF01048094.

[BPR13] S. Boscarino, L. Pareschi, and G. Russo. Implicit-Explicit Runge–Kutta Schemes for Hyperbolic Systems and Kinetic Equations in the Diffusion Limit. SIAM Journal on Scientific Computing, 35(1):A22–A51, 2013. doi:10.1137/110842855.

[BPS12] Stéphane Brull, Vincent Pavan, and Jacques Schneider. Derivation of a BGK model for mixtures. European Journal of Mechanics - B/Fluids, 33:74–86, 2012. doi:10.1016/j.euromechflu.2011.12.003.

[BPW22] G. Bae, M. Pirner, and S. Warnecke. Numerical schemes for multi-species quantum BGK models. manuscript, 2022.

[Bue96] C. Buet. A discrete-velocity scheme for the Boltzmann operator of rarefied gas dynamics. Transport Theory and Statistical Physics, 25(1):33–60, 1996. doi:10.1080/00411459608204829.

[But87] J. C. Butcher. The Numerical Analysis of Ordinary Differential Equations: Runge-Kutta and General Linear Methods. Wiley-Interscience, USA, 1987.

[But16] John Butcher. Numerical Methods for Ordinary Differential Equations. John Wiley & Sons, 08 2016. doi:10.1002/9781119121534.

[BY20] Gi-Chan Bae and Seok-Bae Yun. Quantum BGK Model near a Global Fermi–Dirac Distribution. SIAM Journal on Mathematical Analysis, 52(3):2313–2352, 2020. doi:10.1137/19M1270021.

[BY21] Gi-Chan Bae and Seok-Bae Yun. The Shakhov model near a global Maxwellian, 2021. arXiv:2111.01090.

[CBRY21] Seung Yeon Cho, Sebastiano Boscarino, Giovanni Russo, and Seok-Bae Yun. Conservative semi-Lagrangian schemes for kinetic equations Part II: Applications. Journal of Computational Physics, 436:110281, 2021. doi:10.1016/j.jcp.2021.110281.

[CC70] S. Chapman and T.G. Cowling. The Mathematical Theory of Non-uniform Gases. Cambridge Mathematical Library. Cambridge University Press, Cambridge, 1970.

[CCL12] Anaïs Crestetto, Nicolas Crouseilles, and Mohammed Lemou. Kinetic/fluid micro-macro numerical schemes for Vlasov-Poisson-BGK equation using particles. Kinetic & Related Models, 5(4):787–816, 2012.

[Cer66] Carlo Cercignani. The method of elementary solutions for kinetic models with velocity-dependent collision frequency. Annals of Physics, 40(3):469–481, 1966. doi:10.1016/0003-4916(66)90146-1.

[Cer88] Carlo Cercignani. The Boltzmann Equation and Its Applications, volume 67 of Applied Mathematical Sciences. Springer, New York, 1988. doi:10.1007/978-1-4612-1039-9.

[Cer97] Carlo Cercignani. Temperature, entropy, and kinetic theory. Journal of Statistical Physics, 87(5-6):1097–1109, June 1997. doi:10.1007/BF02181273.

[Cer00] Carlo Cercignani. Rarefied Gas Dynamics: From Basic Concepts to Actual Calculations. Cambridge Texts in Applied Mathematics. Cambridge University Press, Cambridge, 2000.

[CGP12] Yingda Cheng, Irene Gamba, and Jennifer Proft. Positivity-preserving discontinuous Galerkin schemes for linear Vlasov-Boltzmann transport equations. Mathematics of Computation, 81:153–190, 01 2012. doi: 10.1090/S0025-5718-2011-02504-4.

[Cha16] Sydney Chapman. On the law of distribution of molecular velocities, and on the theory of viscosity and thermal conduction, in a non-uniform simple monatomic gas. Philosophical Transactions of the Royal Society of London. Series A, Containing Papers of a Mathematical or Physical Character, 216(538–548):279–348, 1916. doi:10.1098/rsta.1916.0006.

[Cha18] Sydney Chapman. On the kinetic theory of a gas. Part II. — A composite monatomic gas: diffusion, viscosity, and thermal conduction. Philosophical Transactions of the Royal Society of London. Series A, Containing Papers of a Mathematical or Physical Character, 217(549–560):115–197, 1918. doi:10.1098/rsta.1918.0005.

[Chu65] C. K. Chu. Kinetic-Theoretic Description of the Formation of a Shock Wave. The Physics of Fluids, 8(1):12–22, 1965. doi:10.1063/1.1761077.

[CIP94] Carlo Cercignani, Reinhard Illner, and Mario Pulvirenti. The Mathematical Theory of Dilute Gases, volume 309. Springer, 1994. doi:10.1007/978-1-4419-8524-8.

[CKP20] Anaïs Crestetto, Christian Klingenberg, and Marlies Pirner. Kinetic/Fluid Micro-Macro Numerical Scheme for a Two Component Gas Mixture. Multiscale Modeling and Simulation, 18:970–998, 2020.

[Cla20] Michel Claessens. ITER: The Giant Fusion Reactor: Bringing a Sun to Earth. Springer International Publishing, Cham, 2020. doi:10.1007/978-3-030-27581-5.

[CM14] Nicolas Crouseilles and Giovanni Manfredi. Asymptotic preserving schemes for the Wigner–Poisson–BGK equations in the diffusion limit. Computer Physics Communications, 185(2):448–458, 2014. doi:10.1016/j.cpc.2013.06.002.

[CMS10] Nicolas Crouseilles, Michel Mehrenberger, and Eric Sonnendrücker. Conservative semi-Lagrangian schemes for Vlasov equations. Journal of Computational Physics, 229:1927–1953, 2010.

[CP91] F. Coron and B. Perthame. Numerical Passage from Kinetic to Fluid Equations. SIAM Journal on Numerical Analysis, 28(1):26–42, 1991. doi:10.1137/0728002.

[CT05] Thomas M. Cover and Joy A. Thomas. John Wiley & Sons, Ltd, 2005. doi:10.1002/047174882X.

[Dah63] Germund G. Dahlquist. A special stability problem for linear multistep methods. BIT Numerical Mathematics, 3:27–43, 3 1963. doi:10.1007/BF01963532.

[DB94] J. Delcroix and A. Bers. Physique des Plasmas. InterEditions/ CNRS Editions, 1994.

[DDP11] Pierre Degond, Giacomo Dimarco, and Lorenzo Pareschi. The Moment Guided Monte Carlo Method. International Journal for Numerical Methods in Fluids, 67:189–213, 2011.

[Des92] L. Desvillettes. On asymptotics of the Boltzmann equation when the collisions become grazing. Transport Theory and Statistical Physics, 21(3):259–276, 1992. doi:10.1080/00411459208203923.

[Deu20] Deutsche Physikalische Gesellschaft. Komplexität begreifen - Nobelpreis für Physik 2021. Physikkonkret, (60), December 2020.

[DL13] Giacomo Dimarco and Raphaël Loubère. Towards an ultra efficient kinetic scheme. Part I: Basics on the BGK equation. Journal of Computaional Physics, 255:680–698, 2013.

[DLD92] P. Degond and B. Lucquin-Desreux. The Fokker-Planck asymptotics of the Boltzmann collision operator in the Coulomb case. Mathematical Models and Methods in Applied Sciences, 02(02):167–182, 1992. doi:10.1142/S0218202592000119.

[DP11] Giacomo Dimarco and Lorenzo Pareschi. Exponential Runge–Kutta Methods for Stiff Kinetic Equations. SIAM Journal on Numerical Analysis, 49(5):2057–2077, 2011. doi:10.1137/100811052.

[DP13] Giacomo Dimarco and Lorenzo Pareschi. Asymptotic Preserving Implicit-Explicit Runge–Kutta Methods for Nonlinear Kinetic Equations. SIAM Journal on Numerical Analysis, 51(2):1064–1087, 2013. doi:10.1137/12087606X.

[DP14] G. Dimarco and L. Pareschi. Numerical methods for kinetic equations. Acta Numerica, 23:369–520, 2014. doi:10.1017/S0962492914000063.

[DP17] Giacomo Dimarco and Lorenzo Pareschi. Implicit-Explicit Linear Multistep Methods for Stiff Kinetic Equations. SIAM Journal on Numerical Analysis, 55(2):664–690, 2017. doi:10.1137/16M1063824.

[DR11] Robert Denk and Reinhard Racke. Kompendium der ANALYSIS, volume 1. Vieweg+Teubner Verlag|Springer, Wiesbaden, 2011. doi:10.1007/978-3-8348-8184-7.

[DS96] J. E. Dennis and Robert B. Schnabel. Numerical Methods for Unconstrained Optimization and Nonlinear Equations. Society for Industrial and Applied Mathematics, 1996. doi:10.1137/1.9781611971200.

[EE07] P. Ehrenfest and T. Ehrenfest. Begriffliche Grundlagen der Statistischen Auffassung in der Mechanik, chapter 11, pages 773–860. Vieweg+Teubner Verlag, Wiesbaden, 1907. doi:10.1007/978-3-663-16028-1_11.

[EEML60] Paul Ehrenfest, Tatiana Ehrenfest, Michael J. Moravcsik, and R. Bruce Lindsay. The Conceptual Foundations of the Statistical Approach in Mechanics. Physics Today, 13(7):50–52, 1960. doi:10.1063/1.3057042.

[EHY21] Lukas Einkemmer, Jingwei Hu, and Lexing Ying. An efficient dynamical low-rank algorithm for the Boltzmann-BGK equation close to the compressible viscous flow regime. SIAM Journal on Scientific Computing, 43(5):B1057–B1080, 2021.

[Ein19] Lukas Einkemmer. A Low-Rank Algorithm for Weakly Compressible Flow. SIAM Journal on Scientific Computing, 41(5):A2795–A2814, 2019. doi:10.1137/18M1185417.

[EMV03] Miguel Escobedo, Stéphane Mischler, and Manuel Valle. Homogeneous Boltzmann equation in quantum relativistic kinetic theory. Electronic Journal of Differential Equations, Monographs, 01 2003.

[Ens17] David Enskog. doctoralthesis, Uppsala, 1917.

[Ens21] David Enskog. The numerical calculation of phenomena in fairly dense gases. Arkiv för Matematik, Astronomi och Fysik, 16(1):1–60, 1921.

[EP05] Raffaele Esposito and Mario Pulvirenti. Chapter 1. From Particles to Fluids. Handbook of Mathematical Fluid Dynamics, 3, 12 2005. doi:10.1016/S1874-5792(05)80004-7.

[Eul57] Leonhard Euler. Principes généreaux de l'état d'équilibre des fluides. Mémoires de l'académie des sciences de Berlin, 1:217–273, 1757.

[Eva10] Lawrence C. Evans. Partial differential equations. American Mathematical Society, Providence, R.I., 2010.

[FHJ12] Filbet, Francis, Hu, Jingwei, and Jin, Shi. A numerical scheme for the quantum Boltzmann equation with stiff collision terms. ESAIM: M2AN, 46(2):443–463, 2012. doi:10.1051/m2an/2011051.

[FJ10] Francis Filbet and Shi Jin. A class of asymptotic-preserving schemes for kinetic equations and related problems with stiff sources. Journal of Computational Physics, 229:7625–7648, 2010.

[FN28] Ralph Howard Fowler and L. Nordheim. Electron emission in intense electric fields. Proceedings of the Royal Society of London. Series A, Containing Papers of a Mathematical and Physical Character, 119(781):173–181, 1928. doi:10.1098/rspa.1928.0091.

[FSB01] Francis Filbet, Eric Sonnendrücker, and Pierre Bertrand. Conservative numerical schemes for the Vlasov equation. Journal of Computational Physics, 172:166–187, 2001.

[Ful21] Theresa Full. The Semi-Lagrange Numerical Method for the Vlasov-Poisson Equation and Other Kinetic Equations. Master's thesis, Julius-Maximilians-Universität Würzburg, 2021. URL: https://ifm.mathematik.uni-wuerzburg.de/~klingen/ewExternalFiles/TheresaFullMA.pdf.

[GGHS22] Sigal Gottlieb, Zachary J. Grant, Jingwei Hu, and Ruiwen Shu. High Order Strong Stability Preserving MultiDerivative Implicit and IMEX Runge–Kutta Methods with Asymptotic Preserving Properties. SIAM Journal on Numerical Analysis, 60(1):423–449, 2022. doi:10.1137/21M1403175.

[GH14] Irene M. Gamba and Jeffrey R. Haack. A conservative spectral method for the Boltzmann equation with anisotropic scattering and the grazing collisions limit. Journal of Computational Physics, 270:40–57, 2014. doi:10.1016/j.jcp.2014.03.035.

[GHHH17] Irene M Gamba, Jeffrey R Haack, Cory D Hauck, and Jingwei Hu. A fast spectral method for the Boltzmann collision operator with general collision kernels. SIAM Journal on Scientific Computing, 39(4):B658–B674, 2017.

[GIVW10] Xavier Garbet, Y. Idomura, Laurent Villard, and T. Watanabe. Gyrokinetic simulations of turbulent transport. Nuclear Fusion, 50, 04 2010. doi:10.1088/0029-5515/50/4/043002.

[GK56] E. P. Gross and M. Krook. Model for Collision Processes in Gases: Small-Amplitude Oscillations of Charged Two-Component Systems. Physical Review, 102:593–604, May 1956. doi:10.1103/PhysRev.102.593.

[GMM09] Benjamin Graille, Thierry E. Magin, and Marc Massot. Kinetic Theory of Plasmas: Translational Energy. Mathematical Models and Methods in Applied Sciences, 19(04):527–599, 2009. doi:10.1142/S021820250900353X.

[GMS11] M. Groppi, S. Monica, and G. Spiga. A kinetic ellipsoidal BGK model for a binary gas mixture. EPL (Europhysics Letters), 96(6):64002, dec 2011. doi:10.1209/0295-5075/96/64002.

[Gol06] Francois Golse. The Boltzmann Equation and Its Hydrodynamic Limits, volume 2, chapter 3. 12 2006. doi:10.1016/S1874-5717(06)80006-X.

[Gra49] Harold Grad. On the kinetic theory of rarefied gases. Communications on Pure and Applied Mathematics, 2(4):331–407, 1949. doi:10.1002/cpa.3160020403.

[Gra61] Harold Grad. The many faces of entropy. Communications on Pure and Applied Mathematics, 14(3):323–354, 1961. doi:10.1002/cpa.3160140312.

[Gre73] John M. Greene. Improved Bhatnagar-Gross-Krook model of electron-ion collisions. The Physics of Fluids, 16(11):2022–2023, 1973. doi:10.1063/1.1694254.

[GSB89] V. Garzó, A. Santos, and J. J. Brey. A kinetic model for a multicomponent gas. Physics of Fluids A: Fluid Dynamics, 1(2):380–383, 1989. doi:10.1063/1.857458.

[GST01] Sigal Gottlieb, Chi-Wang Shu, and Eitan Tadmor. Strong Stability-Preserving High-Order Time Discretization Methods. SIAM Review, 43(1):89–112, 2001. doi:10.1137/S003614450036757X.

[GT09] Irene M. Gamba and Sri Harsha Tharkabhushanam. Spectral-Lagrangian methods for collisional models of non-equilibrium statistical states. Journal of Computational Physics, 228(6), 4 2009. doi:10.1016/j.jcp.2008.09.033.

[Ham65] Bernard B. Hamel. Kinetic Model for Binary Gas Mixtures. The Physics of Fluids, 8(3):418–425, 1965. doi:10.1063/1.1761239.

[HB13] J.A.F. Hittinger and J.W. Banks. Block-structured adaptive mesh refinement algorithms for Vlasov simulation. Journal of Computational Physics, 241:118–140, 2013. doi:10.1016/j.jcp.2013.01.030.

[HHK$^+$21] J. Haack, C. Hauck, C. Klingenberg, M. Pirner, and S. Warnecke. A Consistent BGK Model with Velocity-Dependent Collision Frequency for Gas Mixtures. Journal of Statistical Physics, 184(3), 9 2021. doi:10.1007/s10955-021-02821-2.

[HHK$^+$22] Jeffrey Haack, Cory Hauck, Christian Klingenberg, Marlies Pirner, and Sandra Warnecke. Numerical schemes for multi-species BGK equations with velocity-dependent collision frequency. submitted, 2022. arXiv:2202.05652.

[HHM17a] Jeffrey R. Haack, Cory D. Hauck, and Michael S. Murillo. A conservative, entropic multispecies BGK model. Journal of Statistical Physics, 168(4):826–856, 2017.

[HHM17b] Jeffrey R. Haack, Cory D. Hauck, and Michael S. Murillo. Interfacial mixing in high-energy-density matter with a multiphysics kinetic model. Physical Review E, 96:063310, Dec 2017. doi:10.1103/PhysRevE.96.063310.

[Hin20] Haye Hinrichsen. Statistical Physics and Thermodynamics - Lecture notes, June 2020.

[HJ11] Jingwei Hu and Shi Jin. On kinetic flux vector splitting schemes for quantum Euler equations. Kinetic & Related Models, 4(2):517–530, 2011.

[HJL17] Jingwei Hu, Shi Jin, and Qin Li. Asymptotic-Preserving Schemes for Multiscale Hyperbolic and Kinetic Equations. Handbook of Numerical Analysis, 18:103–129, 2017.

[Hol66] Lowell H. Holway. New Statistical Models for Kinetic Theory: Methods of Construction. The Physics of Fluids, 9(9):1658–1673, 1966. doi:10.1063/1.1761920.

[HPRV20] Michael Herty, Gabriella Puppo, Sebastiano Roncoroni, and Giuseppe Visconti. The BGK approximation of kinetic models for traffic. Kinetic & Related Models, 13:279–307, 01 2020. doi:10.3934/krm.2020010.

[HPV22] Michael Herty, Gabriella Puppo, and Giuseppe Visconti. Model of vehicle interactions with autonomous cars and its properties. Discrete and Continuous Dynamical Systems - B, 01 2022. doi:10.3934/dcdsb.2022100.

[HSZ18] Jingwei Hu, Ruiwen Shu, and Xiangxiong Zhang. Asymptotic-Preserving and Positivity-Preserving Implicit-Explicit Schemes for the Stiff BGK Equation. SIAM Journal on Numerical Analysis, 56(2):942–973, 2018. doi:10.1137/17M1144362.

[HV03] Willem Hundsdorfer and Jan G. Verwer. Numerical Solution of Time-Dependent Advection-Diffusion-Reaction Equations. Springer, Berlin, Heidelberg, 2003. doi:10.1007/978-3-662-09017-6.

[Ins] Clay Mathematics Institute. Millenium Problems: Navier-Stokes Equation. URL: https://www.claymath.org/millennium-problems/navier%E2%80%93stokes-equation.

[Jin95] Shi Jin. Runge-Kutta Methods for Hyperbolic Conservation Laws with Stiff Relaxation Terms. Journal of Computational Physics, 122:51–67, 1995.

[JL13] Shi Jin and Qin Li. A BGK-penalization-based asymptotic-preserving scheme for the multispecies Boltzmann equation. Numerical Methods for Partial Differential Equations, 29:1056–1080, 05 2013. doi:10.1002/num.21746.

[Joh71] Fritz John. Partial Differential Equations, volume 1 of Applied Mathematical Sciences. Springer, New York, 1971. doi:10.1007/978-1-4615-9966-1.

[JP00] Shi Jin and Lorenzo Pareschi. Discretization of the Multiscale Semiconductor Boltzmann Equation by Diffusive Relaxation Schemes. Journal of Computational Physics, 161:312–330, 2000.

[Jun00] Michael Junk. Maximum entropy for reduced moment problems. Mathematical Models and Methods in Applied Sciences, 10(07):1001–1025, 2000. doi:10.1142/S0218202500000513.

[Kac59] Mark Kac. Probability and Related Topics in Physical Sciences. Interscience Publishers, London, 1959.

[KC03] Christopher A. Kennedy and Mark H. Carpenter. Additive Runge–Kutta schemes for convection–diffusion–reaction equations. Applied Numerical Mathematics, 44(1):139–181, 2003. doi:10.1016/S0168-9274(02)00138-1.

[KHH15] C. Kristopher Garrett, Cory Hauck, and Judith Hill. Optimization and large scale computation of an entropy-based moment closure. Journal of Computational Physics, 302:573–590, December 2015. doi:10.1016/j.jcp.2015.09.008.

[Kir46] John G. Kirkwood. The Statistical Mechanical Theory of Transport Processes I. General Theory. The Journal of Chemical Physics, 14(3):180–201, 1946. doi:10.1063/1.1724117.

[Kir47] John G. Kirkwood. The Statistical Mechanical Theory of Transport Processes II. Transport in Gases. The Journal of Chemical Physics, 15(1):72–76, 1947. doi:10.1063/1.1746292.

[KM91] Katsuhisa Koura and Hiroaki Matsumoto. Variable soft sphere molecular model for inverse-power-law or Lennard-Jones potential. Physics of Fluids A: Fluid Dynamics, 3(10):2459–2465, 1991. doi:10.1063/1.858184.

[KN30] S. Kikuchi and L. Nordheim. Über die kinetische Fundamentalgleichung in der Quantenstatistik. Zeitschrift für Physik A Hadrons and nuclei, 60(9-10):652–662, 1930.

[KP17] Christian Klingenberg and Marlies Pirner. Existence, Uniqueness and Positivity of solutions for BGK models for mixtures. Journal of Differential Equations, 264, 09 2017. doi:10.1016/j.jde.2017.09.019.

[KPP17] Christian Klingenberg, Marlies Pirner, and Gabriella Puppo. A consistent kinetic model for a two-component mixture with an application to plasma. Kinetic & Related Models, 10(2):445–465, 2017.

[KPP18] Christian Klingenberg, Marlies Pirner, and Gabriella Puppo. Kinetic ES-BGK Models for a Multi-component Gas Mixture, pages 195–208. 06 2018. doi:10.1007/978-3-319-91548-7_15.

[KT73] Nicholas A. Krall and Alvin W. Trivelpiece. Principles of Plasma Physics. American Journal of Physics, 41(12):1380–1381, 1973. doi:10.1119/1.1987587.

[Lan37] L. D. Landau. Kinetic equation for the case of Coulomb interaction. Zhurnal Eksperimental'noi i Teoreticheskoi Fiziki, 7, 1937.

[Lan46] Lev Davidovich Landau. On the vibrations of the electronic plasma. Journal of Physics (USSR), 10:25, 1946.

[Lan75] Oscar E. Lanford. Time evolution of large classical systems, chapter 1, pages 1–111. Springer Berlin Heidelberg, Berlin, Heidelberg, 1975. doi:10.1007/3-540-07171-7_1.

[Leb95] Joel L. Lebowitz. Microscopic Reversibility and Macroscopic Behavior: Physical Explanations and Mathematical Derivations, chapter 1, pages 1–20. Springer, Berlin Heidelberg, 1995.

[Len60] Andrew Lenard. On Bogoliubov's kinetic equation for a spatially homogeneous plasma. Annals of Physics, 10(3):390–400, 1960. doi:10.1016/0003-4916(60)90003-8.

[LeV02] Randall J. LeVeque. Finite Volume Methods for Hyperbolic Problems. Cambridge Texts in Applied Mathematics. Cambridge University Press, 2002. doi:10.1017/CBO9780511791253.

[LL69] L. D. Landau and E. M. Lifschitz. Lehrbuch der theoretischen Physik I, Mechanik. Akademie-Verlag, Berlin, 1969.

[LL91] L. D. Landau and E. M. Lifschitz. Lehrbuch der theoretischen Physik VI, Hydrodynamik. Akademie-Verlag, Berlin, 1991.

[LM84] Y. T. Lee and R. M. More. An electron conductivity model for dense plasmas. The Physics of Fluids, 27(5):1273–1286, 1984. doi:10.1063/1.864744.

[LM05] Mohammed Lemou and Luc Mieussens. Implicit Schemes for the Fokker-Planck-Landau Equation. SIAM Journal on Scientific Computing, 27:809–830, 10 2005. doi:10.1137/040609422.

[Low04] Robert B. Lowrie. A comparison of implicit time integration methods for nonlinear relaxation and diffusion. Journal of Computational Physics, 196(2):566–590, 2004. doi:10.1016/j.jcp.2003.11.016.

[LP19] Liu Liu and Marlies Pirner. Hypocoercivity for a BGK model for gas mixtures. Journal of Differential Equations, 267(1):119–149, 2019. doi:10.1016/j.jde.2019.01.006.

[LW60] Peter D. Lax and Burton Wendroff. Systems of conservation laws. Communications on Pure and Applied Mathematics, 13:217–237, 1960.

[Mar21] Simon Markfelder. Convex Integration Applied to the Multi-Dimensional Compressible Euler Equations. Lecture Notes in Mathematics. Springer, Cham, 2021. doi:10.1007/978-3-030-83785-3.

[Max65] James Clerk Maxwell. A dynamical theory of the electromagnetic field. Philosophical Transactions of the Royal Society of London, 155:459–512, 1865. doi:10.1098/rstl.1865.0008.

[Max67] James Clerk Maxwell. IV. On the dynamical theory of gases. Philosophical Transactions of the Royal Society of London, 157:49–88, 1867. doi:10.1098/rstl.1867.0004.

[Mie00] Luc Mieussens. Discrete velocity model and implicit scheme for the BGK equation of rarefied gas dynamic. Mathematical Models and Methods in Applied Sciences, 10, 11 2000. doi:10.1142/S0218202500000562.

[Mie01] L Mieussens. Convergence of a discrete-velocity model for the Boltzmann-BGK equation. Computers & Mathematics with Applications, 41(1):83–96, 2001. doi:10.1016/S0898-1221(01)85008-2.

[MP06] Clément Mouhot and Lorenzo Pareschi. Fast Algorithms for Computing the Boltzmann Collision Operator. Mathematics of Computation, 75(256):1833–1852, 2006.

[MS04] Luc Mieussens and Henning Struchtrup. Numerical comparison of Bhatnagar–Gross–Krook models with proper Prandtl number. Physics of Fluids, 16(8):2797–2813, 2004. doi:10.1063/1.1758217.

[MTH$^+$14] Alessandro Munafo, Erik Torres, Jeffrey Haack, Irene M. Gamba, and Theirry Magin. A Spectral-Lagrangian Boltzmann Solver for a Multi-Energy Level Gas. Journal of Computtional Physics, 264:152–176, 2014.

[MV10] C. Mouhot and C. Villani. Landau damping. Journal of Mathematical Physics, 51:015204–1–015204–7, 2010.

[MV11] C. Mouhot and C. Villani. On Landau damping. Acta Mathematica, 207(1):29–201, 2011. doi:10.1007/s11511-011-0068-9.

[MY12] Bagus Putra Muljadi and Jaw-Yen Yang. Simulation of shock wave diffraction by a square cylinder in gases of arbitrary statistics using a semiclassical Boltzmann-Bhatnagar-Gross-Krook equation solver. Proceedings of the Royal Society A: Mathematical, Physical and Engineering Sciences, 468(2139):651–670, 2012. doi:10.1098/rspa.2011.0275.

[New87] Isaac Newton. Philosophiae Naturalis Principia Mathematica, 1687.

[PB52] David Pines and David Bohm. A Collective Description of Electron Interactions: II. Collective vs Individual Particle Aspects of the Interactions. Physical Review, 85:338–353, Jan 1952. doi:10.1103/PhysRev.85.338.

[Pir18] Marlies Pirner. Kinetic modelling of gas mixtures. doctoralthesis, Würzburg University Press, 2018. doi:10.25972/WUP-978-3-95826-081-8.

[Pir21] Marlies Pirner. A Review on BGK Models for Gas Mixtures of Mono and Polyatomic Molecules. Fluids, 6(11), 2021. doi:10.3390/fluids6110393.

[PP07] Sandra Pieraccini and Gabriella Puppo. Implicit–Explicit Schemes for BGK Kinetic Equations. Journal of Scientific Computing, 32(1):1–28, 2007. doi:10.1007/s10915-006-9116-6.

[PR00] Lorenzo Pareschi and Giovanni Russo. Numerical solution of the Boltzmann equation I: Spectrally accurate approximation of the collision operator. SIAM journal on numerical analysis, 37(4):1217–1245, 2000.

[PR05] Lorenzo Pareschi and Giovanni Russo. Implicit–Explicit Runge–Kutta Schemes and Applications to Hyperbolic Systems with Relaxation. Journal of Scientific Computing, 25(1):129–155, 2005. doi:10.1007/s10915-004-4636-4.

[PR18] S. Pennisi and Tommaso Ruggeri. A New BGK Model for Relativistic Kinetic Theory of Monatomic and Polyatomic Gases. Journal of Physics: Conference Series, 1035:012005, 05 2018. doi:10.1088/1742-6596/1035/1/012005.

[PT21] Gabriella Puppo and Andrea Tosin, editors. Mathematical Descriptions of Traffic Flow: Micro, Macro and Kinetic Models. SEMA SIMAI Springer Series. Springer, Cham, 2021. doi:10.1007/978-3-030-66560-9.

[Pup19] Gabriella Puppo. Kinetic models of BGK type and their numerical integration. In Proceeding of 9th Summer School on 'Methods and Models of Kinetic Theory' (M&MKT 2018) Porto Ercole (Grosseto, Italy) June 10-16, 2018, volume 10, pages 299–349, 2019.

[PWed] Marlies Pirner and Sandra Warnecke. A review on a general multi-species BGK model: modelling, theory and numerics. In Conference Proceedings. The Legacy of Carlo Cercignani: from Kinetic Theory to Turbulence Modeling. Springer, accepted. arXiv:2203.15284.

[QS11] Jing-Mei Qiu and Chi-Wang Shu. Positivity preserving semi-Lagrangian discontinuous Galerkin formulation: Theoretical analysis and application to the Vlasov-Poisson system. Journal of Computational Physics, 230:8386–8409, 2011.

[QSS07] Alfio Quarteroni, Riccardo Sacco, and Fausto Saleri. Numerical Mathematics. Texts in Applied Mathematics. Springer, Berlin Heidelberg, 2007. doi:10.1007/b98885.

[RF09] Giovanni Russo and Francis Filbet. Semilagrangian schemes applied to moving boundary problems for the BGK model of rarefied gas dynamics. Kinetic & Related Models, 2(1):231–250, 2009.

[RMJ57] Marshall N. Rosenbluth, William Michael Macdonald, and David L. Judd. Fokker-Planck Equation for an Inverse-Square Force. Physical Review, 107:1–6, 1957.

[RS07] Tommaso Ruggeri and Srboljub Simić. On the hyperbolic system of a mixture of Eulerian fluids: a comparison between single- and multi-temperature models. Mathematical Methods in the Applied Sciences, 30(7):827–849, 2007. doi:10.1002/mma.813.

[SB02] J. Stoer and R. Bulirsch. Introduction to Numerical Analysis. Texts in Applied Mathematics. Springer, New York, 2002. doi:10.1007/978-0-387-21738-3.

[SBT17] Roman Pascal Schaerer, Pratyuksh Bansal, and Manuel Torrilhon. Efficient algorithms and implementations of entropy-based moment closures for rarefied gases. Journal of Computational Physics, 340:138–159, 2017.

[Sch06] Franz Schwabl. Statistische Mechanik. Springer, Berlin Heidelberg, 2006. doi:10.1007/3-540-31097-5.

[Sch07] Franz Schwabl. Quantum Mechanics. Springer, Berlin Heidelberg, 2007. doi:10.1007/978-3-540-71933-5.

[Sha68] E. M. Shakhov. Generalization of the Krook kinetic relaxation equation. Fluid Dynamics, 3:95–96, 1968.

[Shu88] Chi-Wang Shu. Total-Variation-Diminishing Time Discretizations. SIAM Journal on Scientific and Statistical Computing, 9(6):1073–1084, 1988. doi:10.1137/0909073.

[SJB66] I. P. Shkarofsky, T. W. Johnston, and M. P. Bachynski. The Particle Kinetics of Plasmas. Reading, Mass., Addison-Wesley Pub. Co., 1966.

[SM16] Liam G. Stanton and Michael S. Murillo. Ionic transport in high-energy-density matter. Physical Review E, 93:043203, Apr 2016. doi:10.1103/PhysRevE.93.043203.

[SO88] Chi-Wang Shu and Stanley Osher. Efficient implementation of essentially non-oscillatory shock-capturing schemes. Journal of Computational Physics, 77(2):439–471, 1988. doi:10.1016/0021-9991(88)90177-5.

[Sod78] Gary A Sod. A survey of several finite difference methods for systems of nonlinear hyperbolic conservation laws. Journal of Computational Physics, 27(1):1–31, 1978. doi:10.1016/0021-9991(78)90023-2.

[Son19] E. Sonnendrücker. Numerical Methods for the Vlasov-Maxwell equations. manuscript, 2019.

[Son22] Kerstin Sonnabend. Fusion in neuen Sphären. Physik Journal, 21(3):17, March 2022.

[SR09] Laure Saint-Raymond. Hydrodynamic Limits of the Boltzmann Equation, volume 1971. 01 2009. doi:10.1007/978-3-540-92847-8.

[SRBG99] Eric Sonnendrücker, Jean Roche, Pierre Bertrand, and Alain Ghizzo. The Semi-Lagrangian Method for the Numerical Resolution of the Vlasov Equation. Journal of Computational Physics, 149(2):201–220, 1999. doi:10.1006/jcph.1998.6148.

[SS01] Victor Sofonea and Robert F. Sekerka. BGK models for diffusion in isothermal binary fluid systems. Physica A-statistical Mechanics and Its Applications, 299:494–520, 2001.

[Str68] Gilbert Strang. On the Construction and Comparison of Difference Schemes. SIAM Journal on Numerical Analysis, 5(3):506–517, 1968. doi:10.1137/0705041.

[Str97] Henning Struchtrup. The BGK-model with velocity-dependent collision frequency. Continuum Mechanics and Thermodynamics, 9(1):23–31, 1997. doi:10.1007/s001610050053.

[Str05] Henning Struchtrup. Macroscopic Transport Equations for Rarefied Gas Flows–Approximation Methods in Kinetic Theory. Springer, Berlin, Heidelberg, 01 2005. doi:10.1007/3-540-32386-4.

[Str06] Robert Strain. On the Linearized Balescu-Lenard Equation. Communications in Partial Differential Equations, 32, 03 2006. doi:10.1080/03605300601088609.

[SY08] Yu-Hsin Shi and J.Y. Yang. A gas-kinetic BGK scheme for semiclassical Boltzmann hydrodynamic transport. Journal of Computational Physics, 227(22):9389–9407, 2008. doi:10.1016/j.jcp.2008.06.036.

[TM05] Roger Temam and Alain Miranville. Mathematical Modeling in Continuum Mechanics. Cambridge University Press, 2 edition, 2005. doi:10.1017/CBO9780511755422.

[Tor09] E.F. Toro. Riemann Solvers and Numerical Methods for Fluid Dynamics: A Practical Introduction. Springer, Berlin, Heidelberg, 2009. doi:10.1007/b79761.

[Ueh34] E. A. Uehling. Transport Phenomena in Einstein-Bose and Fermi-Dirac Gases. II. Physical Review, 46:917–929, Nov 1934. doi:10.1103/PhysRev.46.917.

[UU33] E. A. Uehling and G. E. Uhlenbeck. Transport Phenomena in Einstein-Bose and Fermi-Dirac Gases. I. Physical Review, 43:552–561, Apr 1933. doi:10.1103/PhysRev.43.552.

[Vil02] Cédric Villani. A Review of Mathematical Topics in Collisional Kinetic Theory. Handbook of Mathematical Fluid Dynamics, 1, 12 2002. doi:10.1016/S1874-5792(02)80004-0.

[Vil08] Cédric Villani. H-theorem and beyond: Boltzmann's entropy in today's mathematics, Expanded version of my lecture at the Boltzmann memorial in Munich. In Boltzmann's legacy, pages 129–143, Zürich, 2008. ESI Lect. Math. Phys., Eur. Math. Soc.

[WMZ12] Lei Wu, Jianping Meng, and Yong-hao Zhang. Kinetic modelling of the quantum gases in the normal phase. Royal Society of London Proceedings Series A, 468:1799–1823, 06 2012. doi:10.1098/rspa.2011.0673.

[XCX21] Xiaocong Xu, Yipei Chen, and Kun Xu. Modeling and computation for non-equilibrium gas dynamics: Beyond single relaxation time kinetic models. Physics of Fluids, 33(1):011703, 2021. doi:10.1063/5.0036203.

[Xu01] Kun Xu. A Gas-Kinetic BGK Scheme for the Navier–Stokes Equations and Its Connection with Artificial Dissipation and Godunov Method. Journal of Computational Physics, 171(1):289–335, 2001. doi:10.1006/jcph.2001.6790.

[YH09] Jaw-Yen Yang and Li-Hsin Hung. Lattice Uehling-Uhlenbeck Boltzmann-Bhatnagar-Gross-Krook hydrodynamics of quantum gases. Physical Review E, 79:056708, May 2009. doi:10.1103/PhysRevE.79.056708.

[Yvo35] J. Yvon. La Théorie Statistique des Fluides et l'Équation d'Etat, Actualités Scientifiques et Industrialles, volume 203. Hermann, Paris, 1935.

[ZHCe22] A.B. Zylstra, O.A. Hurricane, D.A. Callahan, and et.al. Burning plasma achieved in inertial fusion. Nature, 601:542–548, 2022. doi:10.1038/s41586-021-04281-w.

[ZS05] Yingsong Zheng and Henning Struchtrup. Ellipsoidal statistical Bhatnagar–Gross–Krook model with velocity-dependent collision frequency. Physics of Fluids, 17(12):127103, 2005. doi:10.1063/1.2140710.